HERBAL DRUGS AS THERAPEUTIC AGENTS

Herbal Drugs as Therapeutic Agents

Amritpal Singh Saroya
Herbal Consultant

CRC Press is an imprint of the
Taylor & Francis Group, an **informa** business
A SCIENCE PUBLISHERS BOOK

CRC Press
Taylor & Francis Group
6000 Broken Sound Parkway NW, Suite 300
Boca Raton, FL 33487-2742

First issued in paperback 2018

ISBN-13: 978-1-4665-9860-7 (hbk)
ISBN-13: 978-1-138-37488-1 (pbk)

Library of Congress Cataloging-in-Publication Data

Amritpal Singh, 1971-
Herbal drugs as therapeutic agents / Amritpal Singh Saroya.
pages cm
Includes bibliographical references and index.
ISBN 978-1-4665-9860-7 (hardcover : alk. paper)
1. Herbs--Therapeutic use. I. Title.

RM666.H33.A4948 2014
615.3’21--dc23

2013047361

Visit the Taylor & Francis Web site at
http://www.taylorandfrancis.com

CRC Press Web site at
http://www.crcpress.com

Preface

The chief aim of writing a book titled *Herbal Drugs as Therapeutic Agents* is to highlight the contribution of herbal drugs to pharmacology and in drug discovery. My previous book on *Herbalism, Phytochemistry and Ethnopharmacology* was well received and this motivated me to write a book on therapeutic scope of herbal drugs. Several herbal drugs and isolated constituents have entered clinical trials in recent times and positive outcomes have been reported.

Herbal medicine is going to play a significant role in future healthcare industry. In the past, medicinal plants have provided us with life-saving drugs, particularly in oncology. To name a few—atropine, digoxin, morphine, paclitaxel, pilocarpine, reserpine, scopolamine, topotecan and vincristine. However, several of these compounds have outlived their usefulness in light of better alternatives.

Herbal medicine is interdisciplinary subject and the expert herbal scientist blends traditional herbal medicine with botany, ethnobotany, phytochemistry, pharmacognosy, pharmacology and allopathic medicine. A large part of the world's population depends upon traditional herbal medicine for their daily health requirements, especially in developing countries. Even in industrialized countries the use of plant-based remedies is widespread and numerous pharmaceuticals are based on or derived from plant compounds.

The book starts with a chapter on reported pharmacological activities of withanolides, followed by a chapter targeting anticancer role of withaferin A. The chapter on CAM studies some common anticancer therapies. Succeeding chapters throw light on pharmacological investigations on berberine, protopine, piperine, liriodenine, andrographolide, hypericin, hyperforin and above all, anthrquinones. A chapter has been dedicated to alkaloids from Indian medicinal plants and data on pharmacological investigations.

The chapter on anti-arthritic and anti-acne drugs reviews drugs that are beneficial for treatment of arthritis and acne vulgaris. Pharmacological investigations on Indian nootropic, *Convolvulus plauricaulis* Linn. and anticancer plant,*Viscum album* Linn.- have also been covered.

The author trusts that the present book will meet the long-felt need for a standard book on herbal therapeutics. The author shall be grateful to readers for pointing out any errors that may be there.

Amritpal Singh

List of the Abbreviations

μg	:	a microgram
μM	:	The micrometre
EC50	:	half maximal effective concentration
ED50	:	*In vitro* or *in vivo* dose of drug that produces 50% of its maximum response or effect
Ex vivo	:	Taking place outside a living organism
G2-M	:	cell cycle phase
i.d.	:	Intradermal route of drug administration
i.g.	:	Intragastric route of drug administration
i.l.	:	Intralesional injection
i.m.	:	Intramascular route of drug administration
i.v.	:	Intravenous
IC50	:	The half maximal inhibitory concentration
In vitro	:	Taking place in a test-tube, culture dish or elsewhere outside a living organism
In vivo	:	Taking place in a living organism
IP	:	Intraperitoneal injection
PO	:	Oral (by mouth) route of drug administration
SC	:	Subcutaneous route of drug administration

Contents

Chapter 1

Withanolides—Phytoconstituents with Significant Pharmacological Activities

1.1 INTRODUCTION

About 50 new withanolides have been found in plants, mainly in the roots and leaves, during the year (Table 1.1). Lavie et al., in 1965 studied the basic structure of withanolides. Chemically, they are a group of naturally occurring oxygenated ergostane type steroids (Fig. 1.1), consisting of lactone in the side chain and 2-en-1-one system in ring A.[1]

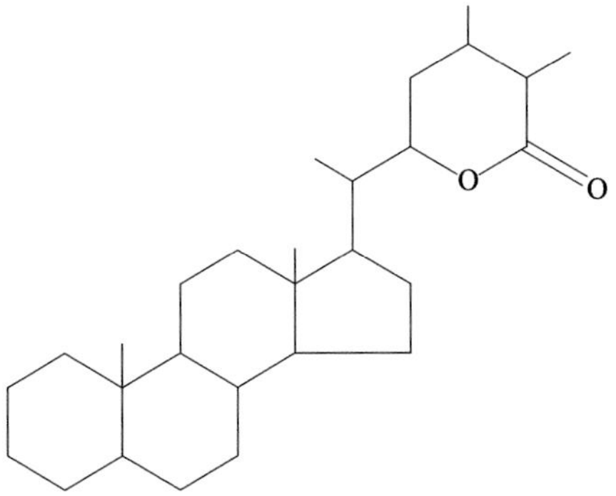

Figure 1.1 Basic withanolide skeleton.

Many are saponins containing an additional acyl group; the rest have glucose at carbon 27.[1] This class of steroid derivative is largely restricted in distribution to the genera *Acnistus, Datura, Discopodium, Dunalia, Jaborosa, Lycium, Nicandra, Physalis, Solanum,* and *Withania,* all belonging to the plant family Solanaceae.[2]

1.2 PREVIOUS REPORTED WORK

Earlier 1α, 3β, 20-trihydroxy witha-5, 24-dienolide and 7α, 27-hydroxy-1-oxo-witha-2, 5-24-trienolide and 7α, 27-dihydroxy-1-oxo- witha-2, 5-24-trienolide were reported from *W. somnifera* chemotype 3.[3] Two minor constituents, 7α, 27-hydroxy-1-oxo-witha-2,5,24-trienolide and 7α, 27-dihydroxy-1-oxo-witha-2,5,24-trienolide have been found in the Indian chemotype of *W. somnifera*.[4] Three withanolides G (Fig. 1.2), H and J, along with 20-hydroxy-1-oxo-witha-2,5,16,24-trienolide have been reported from *W. somnifera* chemotype III.[5-7] Three withanolides I, J and K have been isolated from *W. somnifera* chemotype III.[5] Withanolide U was reported from *W. somnifera* chemotype III.[8]

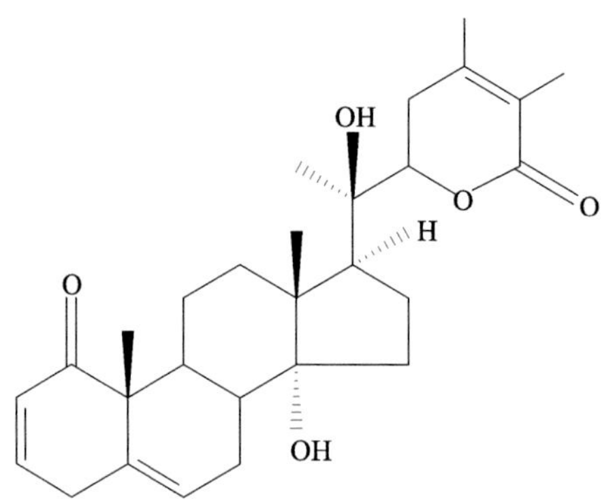

Figure 1.2 Structure of withanolide G.

Withaferin A (Fig. 1.3) was the first compound isolated as a major compound in *W. somnifera* chemotype I.[1] 27-deoxy-withaferin (Fig. 1.4) was also reported to contain withaferin A.[9] Withaferin A is thought to be the primary pharmacological agent present in the roots and leaves of *W. somnifera*.[10,11]

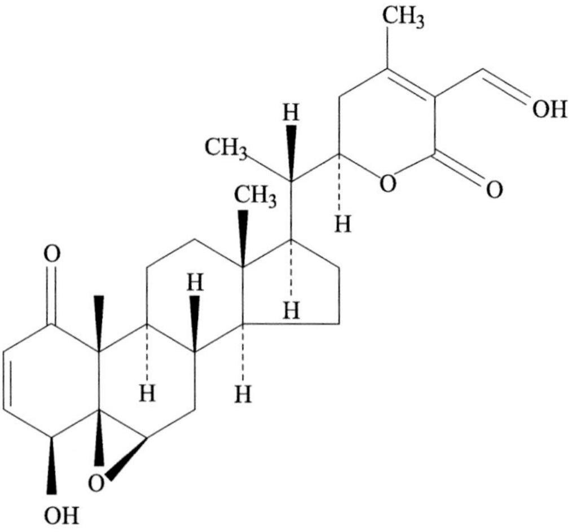

Figure 1.3 Structure of withaferin A.

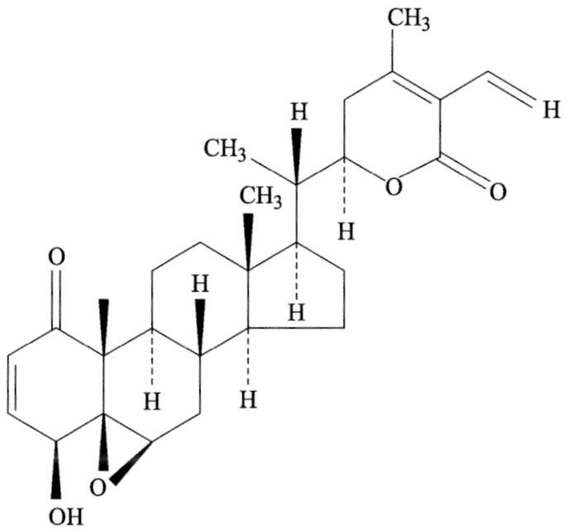

Figure 1.4 Structure of 27-deoxywithaferin A.

Withanolide D (Fig. 1.5) has been reported from *W. somnifera* chemotype II.[12] Several withanolides such as chlorohyrdin II, 27-O-glucosides (sitoinoside IX and X), and withasomidienone (Fig. 1.6) have been characterized from the roots of *W. somnifera*.[13,14] In the Indian chemotype of *W. somnifera*, jaborosalactone A (Fig. 1.7) and withanolide Y have been isolated.[3,15] Two withanolides, Q and R have been reported from the offspring of Indian chemotype 1 and 3 of *W. somnifera*.[16]

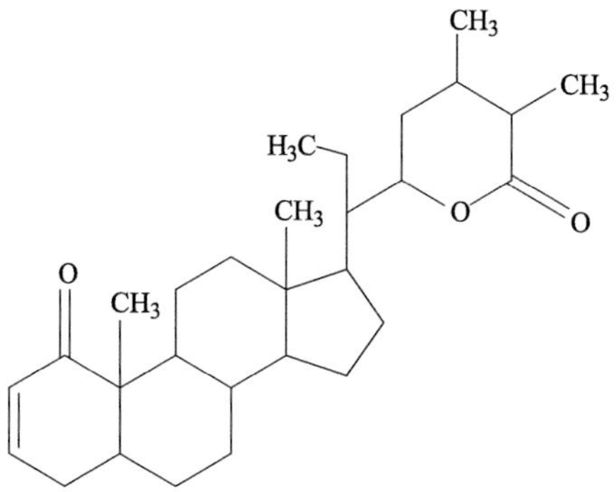

Figure 1.5 Structure of withanolide D.

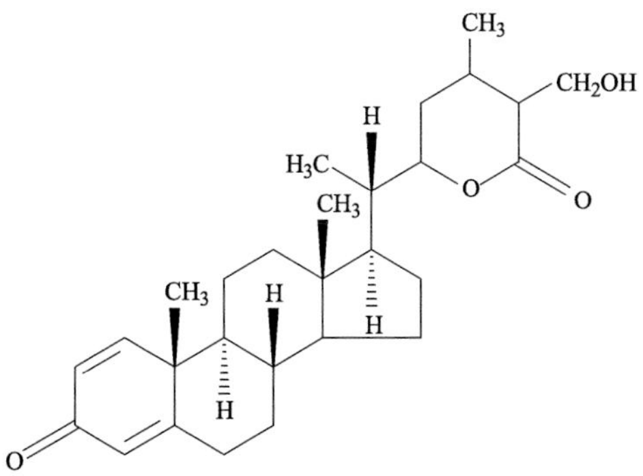

Figure 1.6 Structure of withasomidienone.

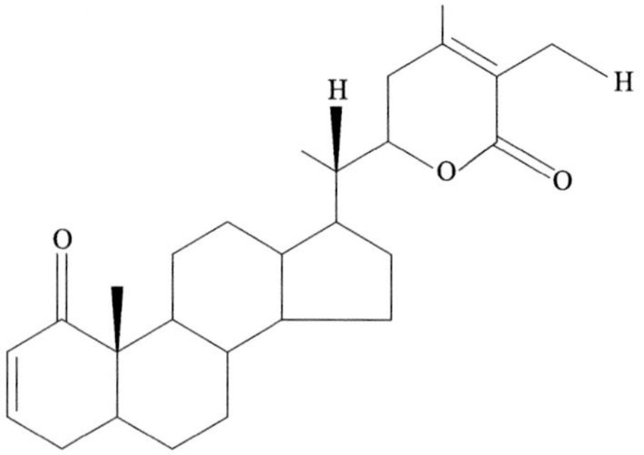

Figure 1.7 Structure of jaborosalactone A.

Withanolide F, E and 4β-hydroxy-withanolide E were isolated from *W. somnifera* chemotype III.[17,18] Furthermore, withanolide S and T have been reported from *W. somnifera* chemotype III.[19] A variety of withanolides including sominolide, soinone, withasomnilide, withasomniferabolide, somniferanolide and somnwithanolide have been reported in the stem bark of *W. somnifera*.[20-24]

1.3 RECENT ADVANCES IN PHARMACOLOGICAL ACTIVITIES

Previous studies reported anti-inflammatory, anti-arthritic, antibiotic, antitumor, immunomodulator and central nervous system effects of withanolides (Table 1.1).[25-37]

Table 1.1 Pre-clinical pharmacological activities of withanolides

Withanolide	*Biological Activity*	*Reference*
Withaferinil	Antitumor	Palyi et al., 1969
Withaferin A	Antitumor	Palyi et al., 1969; Chakrabotri et al., 1974; Ascher et al., 1981; Budhiraja et al., 1987
4β,20-dihydroxy-i-oxo-5β,6β,-epoxy-witha-2, 24-dienolide	Antitumor	Chakrabotri et al., 1974;
Compound WS-1	Hypno-sedative	Kundu et al., 1976
Withanolide -E	Antifeedent	Ascher et al., 1981
Withanolide-5,20α,(*R*) -dihydroxy-6α,7α-epoxy-1-oxo-5α -witha-2, 24-dienolide	Immunomodulator	Bahr et al., 1982
Withanolide-D	Antitumor	Das et al., 1985
3-β-hydroxy-2, 3-dihydro-withanolide	Antibacterial, antitumor, immunomodulator and anti-inflammatory	Budhiraja et al., 1987

Sitoindside-IX, X	Immunomodulator and C.N.S effects	Bhattacharya et al., 1987; Ghosal et al., 1989
Compound WS-2	Antibacterial	Khan et al., 1993
Jaborosalactone R, S and T	-	Bonetto et al., 1995

1.3.1 Antitumor

Withaferin A isolated from the roots of *W. somnifera*, reduced survival of V79 cells in a dose-dependent manner. LD-50 for survival was 16 μM. One-hour treatment with a non-toxic dose of 2.1 μM before irradiation significantly enhanced cell killing, giving a sensitizer enhancement ratio of 1.5 for 37 percent survival and 1.4 for 10 percent survival. Withaferin A induced a G2-M block, with a maximum accumulation of cells in G2-M phase at 4 hr after treatment with 10.5 μM withaferin A in 1 hr.[38]

In another animal study, the alcoholic extract of dried roots of *W. somnifera,* as well withaferin A showed significant antitumor and radiosensitizing effects on experimental tumors *in vivo,* without any noticeable systemic toxicity. Withaferin A gave a sensitizer enhancement ratio of 1.5 for *in vitro* cell killing of V79 Chinese hamster cells at a non-toxic concentration of approximately; 2 μM.[39]

Withaferin A showed marked tumor-inhibitory activity when tested *in vitro* against cells derived from human carcinoma of nasopharynx. It also acted as a mitotic poison arresting the division of cultured human larynx carcinoma cells at metaphase and in HeLa cultures similar to star-metaphase. Withaferin A produced significant retardation of the growth of Ehrlich ascites carcinoma, Sarcoma 180, Sarcoma Black and E 0771 mammary adenocarcinoma in mice in doses of 10, 12, 15 mg/kg body weight.[40]

In this study, it was found that six of the 18 compounds isolated from the methanol extract, enhanced neurite outgrowth in human neuroblastoma SH-SY5Y cells. In withanolide A-treated cells, the length of NF-H-positive processes was significantly increased compared with vehicle-treated cells, whereas, the length of MAP2-positive processes was increased by withanosides IV and VI. The results suggest that axons are predominantly extended by withanolide A, and dendrites by withanosides IV and VI.[41]

Five new withanolides derivatives, were then isolated from the roots of *W. somnifera* together with 14 known compounds. Withanoside VIII, withanoside IX, withanoside XI, withanolide A, withanoside IV, withanoside VI and coagulin showed significant neurite outgrowth at a concentration of 1 μM on a human neuroblastoma SH-SY5Y cell line.[42]

Fifteen new withanolides and two known withanolides, withanolide D and 17α-hydroxywithanolide D, were isolated from the stems, roots and leaves of *Tubocapsicum anomalum* using bioassay-directed fractionation. The majority of withanolides 1, 4-6, 8-10, and 13 showed significant cytotoxic activity against Hep G2, Hep 3B, A-549, MDA-MB-231, MCF-7 and MRC-5 cell lines.[43]

Table 1.2 Withanolides reported during 1996–2009

Withanolide	*Source*	*Reference*
Withaferin A	*Withania somnifera* (Solanaceae)	Devi 1996; Uma et al., 1996; Gupta et al., 1996; Ali et al., 1997; Mohan et al., 2004; Jung et al., 2008. Oh et al., 2008
(20*R*,22*R*)-5α,6β,14α,20,27-pentahydroxy-1-oxowith-24-enolide, (20*S*,22*R*)-5β,6β-epoxy-4β,14β,15α-trihydroxy-1-oxowith-2,24-dienolide, withaphysanolide and viscosalactone B	*Physalis peruviana* (Solanaceae)	Ahmad et al., 1999
Ajugin C (=(20*R*,22*R*)-4β,14α,20,27-tetrahydroxy-1-oxoergosta-2,5,24-trieno-26,22-lactone; and ajugin D (=(20*R*,22*R*)-8β,14α,17β,20,27-pentahydroxy-1-oxoergosta-5,24-dieno-26,22-lactone	*Ajuga parviflora* (Lamiaceae)	Khan et al., 1999
Withanoside VIII, withanoside IX, withanoside XI, withanolide A, withanoside IV, withanoside VI, and coagulin	*Withania somnifera* (Solanaceae)	Zhao et al., 2002
Withanolide A, withanosides IV and VI	*Withania somnifera* (Solanaceae)	Kuboyama et al., 2002
Physagulin D (1→6)-β-D-glucopyranosyl-(1-4)-β-D-glucopyranoside), 27-O-β-D-glucopyranosyl physagulin D, 27-O-β-D-glucopyranosyl viscosalactone B, 4,16-dihydroxy-5β, 6β-epoxyphysagulin D, 4-(1-hydroxy-2,2-dimethylcyclo-propanone)-2,3-dihydrowithaferin A, withaferin A, 2,3-dihydrowithaferin A, viscosalactone B, 27-desoxy-24, 25-dihydrowithaferin A, sitoindoside IX, physagulin D, and withanoside IV	*Withania somnifera* (Solanaceae)	Jayaprakasam and Nair, 2003
Ixocarpalactone A, 2,3-dihydro-3β- methoxyixocarpalactone A, 2, 3-dihydro-3β-methoxyixocarpalactone B, 2,3-dihydroixocarpalactone B, and 4β,7β,20*R*-trihydroxy-1-oxowitha-2,5-dien-22,26-olide	*Physalis philadelphica* (Solanaceae)	Gu et al., 2003
Subtrifloralactones A-E,F-L, new C-18 oxygenated withanolide, 13β-hydroxymethylsubtrifloralactone E and philadelphicalactone A	*Deprea subtriflora* (Solanaceae)	Kinghorn et al., 2003, Su et al., 2003

20β-hydroxy-1-oxo-(22*R*)-witha-2,5,24-trienolide, withacoagulin and a known withanolide, 17β-hydroxy-14α, 20α-epoxy-1-oxo-(22*R*)-witha-3,5,24-trienolide	*Withania coagulans* (Solanaceae)	Rahman et al., 2003
Bracteosin A (=(22*R*)-5β,6α:22,26-diepoxy-4β,28-dihydroxy-3β-methoxyergost-24-ene-1,26-dione), bracteosin B (=(22*R*)-5β,6β: 22,26-diepoxy-4β,28-dihydroxy-3β a-methoxy-1,26-dioxoergost-24-en-19-oic acid), and bracteosin C (=(22*R*)-22,26-epoxy-4β,6β,27-trihydroxy-3β- methoxyergost-24-ene-1,26-dione)	*Ajuga bracteosa* (Lamiaceae)	Naheed et al., 2004
Withanolide A, withanoside IV, and withanoside VI	*Withania somnifera* (Solanaceae)	Tohda et al., 2005
Withanolide 1 and 2	*Jaborosa caulescens* (*var. caulescens* and *var. bipinnatifida*)	Nictota et al., 2005
Daturametelins C, D, E, F and G	*Datura metel* (Solanaceae)	Shingu et al., 2005
Witharifeen (11.α, 12.β.-dihydroxy (20*R*, 22*R*)-21, 24-epoxy-1-oxowitha-2, 5, 25(27)-trien-22, 26-olide) and daturalicin (20*R*, 22*R*)-5.β., 6.β. -14.α.,15.α. -21,24-triepoxy-1-oxowitha-2,25(27)-dien-22,26-olide)	*Datura innoxia* (Solanaceae)	Siddique et al., 2005
Withanone, 27-hydroxy withanolide A, two new withanolides, isowithanone and 6 á,7 á-epoxy-1 á,3β,5 á-trihydroxy-witha-24-enolide	*Withania somnifera* (Solanaceae)	Lal et al., 2006
Cilistepoxide, and cilistadiol	*Solanum sisymbifolium* (Solanaceae)	Niero et al., 2006
Ashwagandhanolide	*Withania somnifera* (Solanaceae)	Subbaraju, 2006

Contd.

Withanolide	*Source*	*Reference*
Daturametelins H-J, daturataturin A and 7,27-dihydroxy-1-oxowitha-2,5, 24-trienolide	*Datura metel* (Solanaceae)	Ma et al., 2006
Fifteen new withanolides, withanolide D and 17α-hydroxywithanolide D	*Tubocapsicum anomalum*	Hsieh et al., 2007
Withanolide Z	*Withania somnifera* (Solanaceae)	Pramanick et al., 2008
Withangulatin A (1) and withangulatin I	*Physalis angulata* (Solanaceae)	Lee et al., 2008
Physacoztolides A-E	*Physalis coztomatl* (Solanaceae)	Perez-Castorena et al., 2006
Withanolide Z	*Withania somnifera* (Solanaceae)	Pramanick et al., 2008
Coagulanolide and withanolides 1-3 and 5	*Withania coagulans* (Solanaceae)	Maurya et al., 2008
Withaferin A and witharistatin	*Withania aristata* (Solanaceae)	Benjumea et al., 2009
Withanolide sulfoxide	*Withania somnifera* (Solanaceae)	Vanisree et al., 2009
12β-acetoxy-4-deoxy-5,6-deoxy-Δ^5–withanolide D and Withanolide D	*Acnistus arborescens*	Codero et al., 2009
Withacoagulins A-F	*Withania coagulens* (Solanaceae)	Huang et al., 2009
Withanoside IV , Withanoside VI, Physagulin D and Withastraronolide	*Withania somnifera* (Solanaceae)	Ahuja et al., 2009

Three new withanolide glycosides called daturametelins (Fig. 1.8) H-J, together with two known ones, daturataturin A and 7,27-dihydroxy-1-oxowitha-2,5,24-trienolide, isolated from the MeOH extract of the aerial parts of *Datura metel* were tested for their antiproliferative activity towards the human colorectal carcinoma (HCT-116) cell line. 7,27-dihydroxy-1-oxowitha-2,5,24-trienolide exhibited the highest activity of the tested withanolides, with an IC50 value of 3.2+/−0.2 μM.[44]

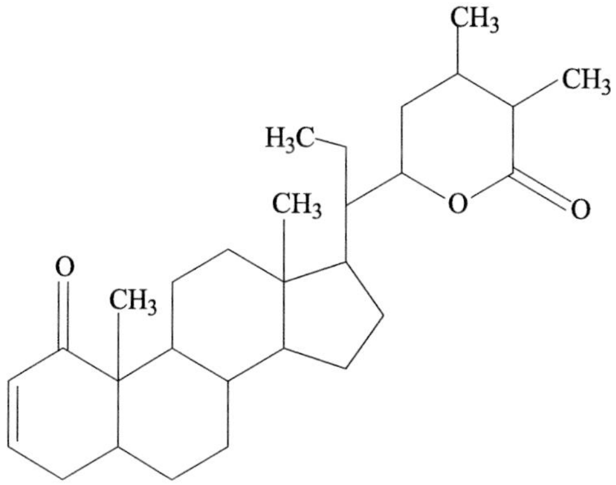

CH_3 CH_3 H_3C CH_3 O O O CH_3

Figure 1.8 Structure of daturametelin.

Extracts of *Withania adpressa*, were tested for their cytotoxicity towards a panel of cancer cell lines (Hep2, HT29, RD, Vero and MDCK), using the (3-[4,5-dimethylthiazol-2-yl]-2,5-diphenyltetrazolium bromide). The bioassay-guided fractionation of the plant extracts resulted in isolation of a novel withanolide 14*α*,15*α*,17*β*,20*β*-tetrahydroxy-1-oxo-(22*R*)-witha-2,5, 24-trienolide and the already identified withanolides F and J. Extract, semi-purified fractions and pure compounds exhibit potent cytotoxicity against human cancer cell lines tested, in a dose-dependent manner.[45]

Four new withanolides, physagulins L-O with seven known withanolides, were isolated from the MeOH extract of the aerial parts of *Physalis angulata*. All eleven compounds were tested for their antiproliferative activities towards human colorectal-carcinoma (HCT-116) and human non-small-cell lung-cancer (NCI-H460) cells. The fifth withanolide compound exhibited the highest anticancer activity against the HCT-116 cell line, with an IC50 value of 1.64+/−0.06 μM. The nineth withanolide compound exhibited the highest cytotoxicity towards the NCI-H460 cell line, with an IC50 value of 0.43+/−0.02 μM.[46]

Withangulatin A and withangulatin I, from the whole plant of *Physalis angulata*, were tested for their cytotoxic activities against two human cancer cell lines, colorectal carcinoma COLO 205 and gastric carcinoma AGS, *in vitro*. Compounds 1 and 2 exhibited inhibitory activities against these two human cancer cells with IC50 values of 16.6 and 1.8 and 53.6 and 65.4 μM, respectively.[47]

In the present study, we demonstrate that a major component of i-Extract and withanone (Fig. 1.9) (i-Factor) protects the normal human fibroblasts against toxicity caused by withaferin A. It increases the *in vitro* division potential of normal human cells that appear to be mediated by decreased accumulation of molecular damage, down regulation of the senescence-specific *β*-galactosidase activity and the senescence marker protein, $p21^{WAF\text{-}1}$, protection against oxidative damage, and induction of proteasomal activity. From these findings it was concluded that i-Extract and withanone, both have anticancer and anti-aging activities. It further points to the molecular link between aging and cancer.[48]

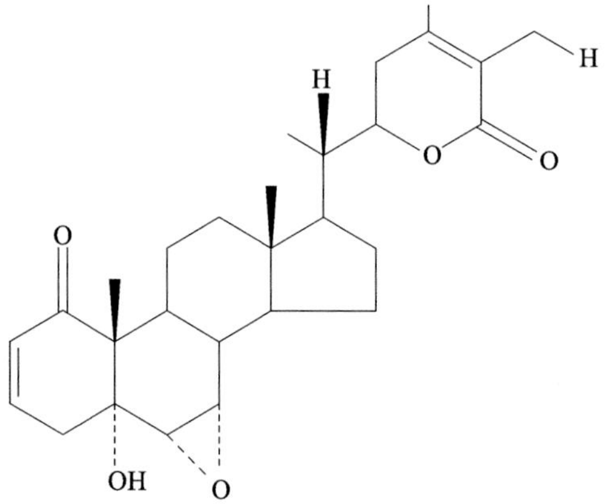

Figure 1.9 Structure of withanone.

A new withanolide, 12*β*-acetoxy-4-deoxy-5,6-deoxy-Δ^5–withanolide D and withanolide D, were isolated from the leaves of *Acnistus arborescens*. Cytotoxic activity of these two compounds against human tumor cell lines HT-29, MCF-7, MKN-45, HEp-2, HeLa, U-937 and two human normal fibroblast cultures, Fib04 and Fib05 were studied. Withanolide D presented *in vitro* cytotoxic activity against tumor cell lines at the low micromolar range (LC50:1.0 to 1.69 μM) and showed a slightly lower activity against Fib04, suggesting moderated selectivity among tumoral and normal cells. No cytotoxic effect was observed for 12*β*-acetoxy-4-deoxy-5,6-deoxy-Δ^5-withanolide D.[49]

The chemokine receptor CCR7 is important for lymphatic invasion of cancer cells and is over expressed in metastatic breast cancer cells. A bioactive withanolide, tubocapsanolide suppressed NF-κB-mediated CCR7 expression in breast cancer cells and attenuated their migration toward lymphatic endothelial cells.[50]

1.3.2 Anti-inflammatory and Anti-oxidant

Raw 264.7 cells stimulated with lipopolysaccharide (LPS) to mimic inflammation, withaferin A inhibited LPS-induced expression of both iNOS protein and mRNA in a dose-dependent manner. To investigate the mechanism by which withaferin A inhibits iNOS gene expression, the researchers examined activation of mitogen-activated protein kinases

(MAPKs) and Akt in Raw 264.7 cells. Withaferin A prevented IκB phosphorylation, blocking the subsequent nuclear translocation of nuclear factor-κB (NF-κB) and inhibiting its DNA binding activity. Moreover, LPS-induced NO production and NF-κB activation were inhibited by SH-6, a specific inhibitor of Akt. The results suggest that withaferin A inhibited inflammation through inhibition of NO production and iNOS expression, by blocking Akt and subsequently downregulating NF-κB activity.[51]

Four novel withanolide glycosides and a withanolide (physagulin D (1→6)-*β*-D-glucopyranosyl-(1-4)-*β*-D-glucopyranoside, 27-O-*β*-D glucopyranosyl physagulin D, 27-O-*β*-D-glucopyranosyl viscosalactone B, 4,16-dihydroxy-5*β*, 6*β*-epoxyphysagulin D and 4-(1-hydroxy-2,2-dimethylcyclo-propanone)-2,3-dihydrowithaferin A) were isolated from the leaves of *W. somnifera.* In addition, seven known withanolides (withaferin A, 2,3-dihydrowithaferin A, viscosalactone B, 27-desoxy-24, 25-dihydrowithaferin A, sitoindoside IX, physagulin D and withanoside IV were isolated. These withanolides were assayed to determine their ability to inhibit cycloxygenase-1 and cyclooxygenase-2 enzymes and lipid peroxidation. The withanolides tested, except for 27-desoxy-24,25-dihydrowithaferin A, showed selective cyclooxygenase-2 enzyme inhibition ranging from 9 to 40 percent at 100 μg/ml. 4,16-dihydroxy-5*β*, 6*β*-epoxyphysagulin D, sitoindoside IX and physagulin D also inhibited lipid peroxidation by 40, 44 and 55 percent, respectively.[52]

Investigation of the methanol extract of *W. somnifera* roots for bioactive constituents yielded a novel withanolide sulfoxide compound along with a known withanolide dimer ashwagandhanolide. Compound 1 was highly selective in inhibiting cyclooxygenase-2 (COX-2) enzyme by 60 percent at 100 μM with no activity against COX-1 enzyme. The IC50 values of compound 1 against human gastric (AGS), breast (MCF-7), central nervous system (SF-268) and colon (HCT-116) cancer cell lines were in the range 0.74–3.63 μM.[53]

1.3.3 Anti-cholinesterase

Three new withanolides, bracteosin A (=(22*R*)-5*β*,6*β*:22,26-diepoxy 4*β*, 28-dihydroxy-3*β*-methoxyergost-24-ene-1,26-dione;), bracteosin B (=(22*R*)5*β*, 6*β*: 22, 26-diepoxy-4*β*,28-dihydroxy-3*β*-methoxy-1,26-dioxoergost-24-en-19 oic acid, and bracteosin C (=(22*R*)-22,26-epoxy- 4*β*,6*β*,27-trihydroxy-3*β*-methoxyergost-24-ene-1,26-dione), and dihy-droclerodin-1, clerodinin A, lupulin A and dihydroajugapitin were isolated from the whole plants of *Ajuga bracteosa*. Bracteosin A-C, exhibited evident inhibitory potential against cholinesterase enzymes in a concentration-dependent manner.[54]

1.3.4 Neuroprotective

In one study, researchers screened the neurite outgrowth activity of herbal drugs, and identified several active constituents. In each compound, neurite outgrowth activity was investigated under amyloid-beta-induced neuritic atrophy. Most of the compounds with neurite regenerative activity also demonstrated memory improvement activity in Alzheimer's disease-model mice. Withanolide derivatives (withanolide A, withanoside IV and withanoside VI) isolated from *W. somnifera*, also showed neurite extension in normal and damaged cortical neurons.[55]

In another animal study, it was investigated, whether withanolide A, isolated from the root of *W. somnifera*, could regenerate neurites and reconstruct synapses in severely damaged neurons. Furthermore the effect of withanolide A on memory-deficient mice showing neuronal atrophy and synaptic loss in the brain was also investigated. Subsequent treatment with withanolide A, induced significant regeneration of both axons and dendrites, in addition to the reconstruction of pre- and post-synapses in the neurons. Withanolide A recovered A β (25-35)-induced memory deficit in mice. At which time, the decline of axons, dendrites and synapses in the cerebral cortex and hippocampus was almost recovered.[56]

1.3.5 Angiogenesis Inhibitor

In an endothelial cell-sprouting assay, it was demonstrated that withaferin A inhibits human umbilical vein endothelial cell (HUVEC) sprouting in three-dimensional collagen-I matrix at doses which are relevant to NF-κB-inhibitory activity. Withaferin A inhibits cell proliferation in HUVECs (IC50 = 12 μM) at doses that are significantly lower than those required for tumor cell lines through a process associated with inhibition of cyclin D1 expression. We propose that the inhibition of NF-κB by withaferin A in HUVECs occurs by interference with the ubiquitin-mediated proteasome pathway as suggested by the increased levels of poly-ubiquitinated proteins. Withaferin A was shown to exert potent anti-angiogenic activity *in vivo* at doses that are 500-fold lower than those previously reported to exert antitumor activity *in vivo*.[57]

1.3.6 Diuretic

Four *Withania aristata* extracts at 100 mg/kg were orally administered to laboratory animals to evaluate their diuretic activity. Two withanolides were isolated from the most active fraction. Both and a mixture of them at 5 and 10 mg/kg were also analyzed as diuretics. Water excretion rate and content of Na (+) and K (+) electrolytes were measured in the urine of saline-loaded animals. *W. aristata* water fraction, the two withanolides and the mixture of these compounds displayed high diuretic activity, with a significant excretion of sodium and potassium ions in laboratory animals. The activity was ascribed to withaferin A and witharistatin.[58]

1.3.7 Hypoglycemic

A new withanolide, coagulanolide along with four known withanolides 1-3 and 5 have been isolated from *Withania coagulans* fruits and their structures were elucidated by spectroscopic techniques. All the compounds showed significant inhibition on postprandial rise in hyperglycemia post-sucrose load in normoglycemic rats as well as streptozotocin-induced diabetic rats. Withanolide 5 showed a significant fall on fasting blood glucose profile and improved glucose tolerance of db/db mice. The db/db mouse is a model of obesity, diabetes, and dyslipidemia. Furthermore Withanolide 5 showed antidyslipidemic activity in db/db mice.[59]

1.3.8 Immunosuppression

Six new withanolides, withacoagulins A-F (1-6), together with 10 known withanolides, 7-16, were isolated from the aerial parts of *Withania coagulans*. These compounds, including the crude extracts of this herb, exhibited strong inhibitory activities on the T- and B-cell proliferation.[60]

1.3.9 Miscellaneous

Two new withanolides, (20R,22R)-5α,β,14α,20,27-pentahydroxy-1-oxowith-24-enolide and (20S,22R)-5β,6β-epoxy-4β,14β,15α-trihydroxy-1-oxowith-2, 24-dienolide, in addition to the known withanolides, withaphysanolide and viscosalactone B were isolated from the whole plant material of *Physalis peruviana*.[61]

The phytochemical study of two species of *Jaborosa caulescens* (*var. caulescens*and *var. bipinnatifida*) yielded four new withanolides 1-4.[62] The whole plant extract of *Deprea subtriflora* yielded, subtrifloralactones A–E and F–L, a new C-18 oxygenated withanolide, 13β-hydroxymethylsubtrifloralactone E (Fig. 1.10), a new α-ionone derivative, (+)-7α,8α-epoxyblumenol B and philadelphicalactone A (Fig. 1.11).[63,64]

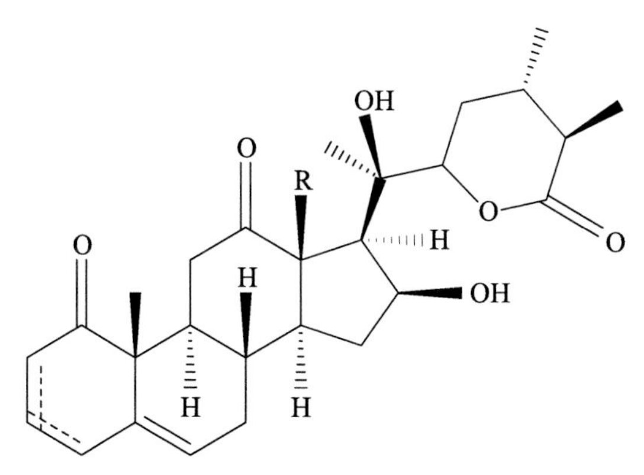

Figure 1.10 Structure of 13β-hydroxymethylsubtrifloralactone E.

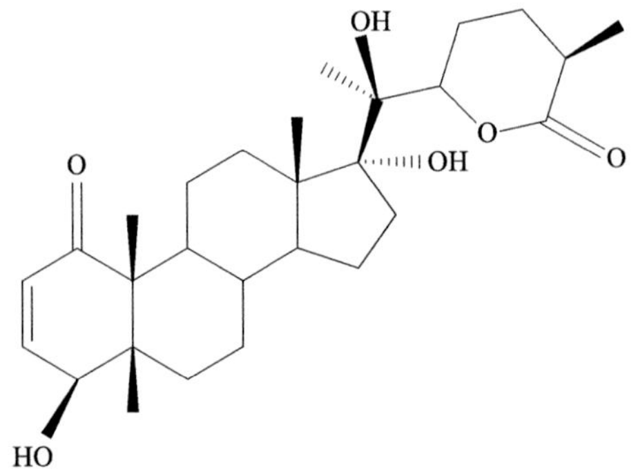

Figure 1.11 Structure of philadelphicalactone A.

Two new withanolides ajugin C (=(20*R*,22*R*)-4*β*,14*α*,20,27-tetrahydroxy-1-oxoergosta-2,5,24-trieno-26,22-lactone and ajugin D (=(20*R*,22*R*)-8*β*,14*α*, 17*β*,20,27-pentahydroxy-1-oxoergosta-5,24-dieno-26,22-lactone were isolated from the whole plant of *Ajuga parviflora*.[65]

Two withanolides, 20*β*-hydroxy-1-oxo-(22*R*)-witha-2,5,24-trienolide and withacoagulin, along with a known withanolide, 17*β*-hydroxy-14*α*, 20*α*-epoxy-1-oxo-(22*R*)-witha-3,5,24-trienolide were isolated from *Withania coagulans*.[66]

Minor new withanolides, daturametelins C, D, E, F and G-Ac were isolated from the methanolic extract of the fresh aerial parts of *Datura metel*. Daturametelins E and F are the first withanolides having a 1-one-3*β*-O-sulfate structure in ring A.[67]

Cilistepoxide and cilistadiol, two new withanolides have been isolated from *Solanum sisymblifolium*.[68] Aerial parts of *Physalis coztomatl* produced a new labdane diterpene, physacoztomatin and five new withanolides, physacoztolides A-E (5-9).[69]

Bioactivity-guided search for novel, plant-derived cancer chemopreventive agents, yielded ixocarpalactone A (Fig. 1.12) and minor new withanolides, 2,3-dihydro-3*β*-methoxyixocarpalactone A, 2,3-dihydro-3*β*-methoxyixocarpalactone B, 2,3-dihydroixocarpalactone B, and 4*β*,7*β*,20*R*-trihydroxy-1-oxowitha-2,5-dien-22,26-olide from the leaves and stems of *Physalis philadelphica*.[70]

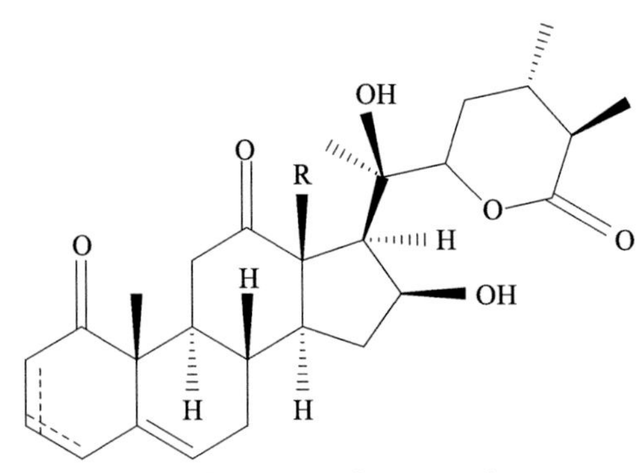

Figure 1.12 Structure of ixocarpalactone A.

The methanolic extract of the aerial parts of *Datura innoxia* produced two new withanolides namely witharifeen (11,12β-dihydroxy (20*R*,22*R*)-21,24-epoxy-1-oxowitha-2, 5, 25(27)-trien-22,26-olide) and daturalicin ((20*R*, 22*R*)-5β,6β-14.é,15α-21,24-triepoxy-1-oxowitha-2,25(27)-dien-22,26-olide).[71]

The chloroform extract of the fresh berries of *Withania somnifera* has been investigated to produce stigmasterol, its glucoside, withanone, 27-hydroxy withanolide A along with two new withanolides, namely, iso-withanone and 6α,7α epoxy-1β 3β,5β trihydroxy-witha-24-enolide.[72]

A new dimeric withanolide, ashwagandhanolide, was isolated from the roots of *W. somnifera*. It displayed growth against human gastric (AGS), breast (MCF-7), central nervous system (SF-268), colon (HCT-116) and lung (Ncl h460) cancer cell lines, with IC50 values in the range 0.43–1.48 µg.[73] n-butanol fraction of the methanolic extract of leaves of *W. somnifera* leaves produced a novel chlorinated withanolide, withanolide Z, along with known withanolides, withanolide B, withanolide A, 27-hydroxywithanolide B and withaferin A.[74]

Recently four glycowithanolides viz. withanoside IV, withanoside VI, physagulin D and withastraronolide, were characterized from multiple shoot cultures of selected accessions AGB002 and AGB025 of *W. somnifera*.[75]

REFERENCES

1. Lavie D, Glotter E, Shvo Y. Constituents of *Withania somnifera*-III—The side chain of Withaferin A. *J Org Chem* 1965; 30: 1774-1778.
2. Kamernitskii AV, Reshetova IR, Krivoruchko VA. Withanolides—A new type of phytosteroids. *Soedin* 1977; 2: 156–186.
3. Vande VV, Lavid D. New Withanoloids of biogenic interest from *Withania somnifera*. *Phytochemistry* 1981; 20: 1359-1364.
4. Kirson I, Glotter E, Lavis D, *et al.*, Constituents of *Withania somnifera* Dunal XII—The Withanolides of an Indian Chemotype. *J Chem Soc* 1971; 52: 2032-2044.
5. Glotter E, Kirson I, Abraham A, *et al.*, Constituents of *Withania somnifera* (Dunal) XIII—The Withanolides of Chemotype III. *Tetrahedron Lett* 1973; 29: 1353-1364.
6. Kirson I, Glotter E. 14 a-hydroxy steroids from *Withania somnifera* (L) Dunal. *J Chem Res Synop* 1980; 10: 338-339.
7. Vande VV, Lavie D. A D16—Withanolide in *Withania somnifera* as a possible precursor for a-side chain. *Phytochemistry* 1982; 21: 731-733.
8. Lavie D, Kirson I, Abraham A. Analysis of hybrids of *Withania somnifera* L. (Dun.) Chemotypes III (Israel) by Indian I (Delhi). *Israle J Chem* 1977; 16: 20-24.
9. Bharat L, Mors WB, Kirson I. *et al.*, 27-deoxy-withaferin from *W. somnifera*. *An Acad Bras Ciec* 1970; 42: 401.
10. Ali M, Shuaib M. Withanolides from the stem bark of *Withania somnifera*. *Phytochemistry* 1997; 44: 1163-1168.

11. Gupta AP, Verma RK, Misra HO, *et al.*, Quantitative determination of Withaferin A in different plant parts of *Withania somnifera* by TLC densitometry. *J Med Arom Plant Sci* 1996; 18: 788-790.
12. Lavie D, Glotter E, Kirson I, *et al.*, Isolation of Withanolide D from *W. somnifera* chemotype II. *Phytochemistry* 1975; 14: 189.
13. Ghosal S, Lal J, Srivastava RS. Immunomodulatory and CNS effect of Sitoindosides IX and X, two regio glycowithanolides from *Withania somnifera. Ind J Nat Prod* 1988; 4: 12-13.
14. Sharma NS, Gottieb HE, Kirson I. Characterization of withasomidienone from roots of *W. somnifera* 12th IUPAC Symposium, *Chem Nat Prod Tenerife*, Spain, 1980.
15. Bessalle R, Lavie D, Froloe F. Withanaloid Y, a withanaloid from a hybrid *Withania somnifera. Phytochemistry* 1987; 26: 1797-1800.
16. Kirson I, Cohen A, Abraham A. Withanolides Q and R, two new 23-hydroxy-steroidal lactones. *J Chem Soc Perkin Trans* 1975; 1: 2136-2138.
17. Lavie D, Glotter E, Kirson I, *et al.*, Steroidal lactones of *Withania somnifera. Harokaech Haivri* 1972; 14: 362-368.
18. Glotter E, Kirson I, Abraham A. Constituents of *Withania somnifera* Dunal XII. The withanolides of an Indian Chemotype. *J Chem Soc* 1971; 1: 2032-2044.
19. Sharma NS, Vande VV, Frolow F, *et al.*, Chlorinated Withanolides from *W. somnifera* and *Acnistus breviflorus. Phytochemistry* 1981; 20: 2547-2552.
20. Rehman A, Jamal SA, Choudhry MI. Withasomnilide and withasomniferabolide from stem=-bark of *W. somnifera. Heterocycl Comp* 1992; 34, 689.
21. Rehman A, Jamal SA, Choudhry MI, *et al.*, Two Withanolides from *Withania somnifera. Phytochemistry* 1991; 30: 3824-3825.
22. Bessalle R, Lavie D. Withanolide C, A chlorinated Withanolide from *Withania somnifera. Phytochemistry* 1992; 31: 3648-3651.
23. Rehman A, Jamal SA, Choudhry MI, *et al.*, New Withanolides from *Withania* spp. *J Nat Prod* 1993; 56: 1000-1006.
24. Ali M, Shuaib M, Ansari SH. Withanolides from the stem bark of *Withania somnifera. Phytochemistry* 1997; 44: 1163-1168.
25. Sethi PD, Subramanian SS. Steroidal lactone of *Withania ashwagandha. Ind J Pharmacol* 1971; 33: 25-26.
26. Kohlmnezer S, Krupniska J. Isolation of an antibiotic substance (unsaturated lactine, m.p. 159-60°C) from the leaves of *Withania somnifera. Disseratones Pharm* 1963; 14: 501-506.
27. Sohat B, Gitter E, Abraham A, *et al.*, Antitumor activity of withaferin A. *Can Chemother Rep* 1967; 51: 51271-51276.
28. Chakrabarti SK, Barun DK, Bandyopadhyay P. Variations in the antitumour constituents of *Withania somnifera* Dunal. *Experienta* 1974; 30: 852-853.

29. Pabji L, Tyihak E, Palyi V. Cytological effects of compounds (Withaferin A and Withferinil) isolated from *Withania somnifera* Dun. *Herba Hung* 1969; 8: 73.
30. Budhiraja RD, Sudhir S. Review of biological activity of Withenolides (Antibacterial Antitumor, Immunomodulating, Antiinflammatory and insect antifeedcent). *J Sci Ind Res* 1983; 46: 488-491.
31. Bahr V, Hansel R. Immunomodulatory properties of 5, 20-a(r) dihydroxy-6a,7a,epoxy-1-oxo5a, with2, 2g, dienolide and solasodine. *Planta Med* 1982; 44: 32-38.
32. Das H, Dultan SK, Bhattacharya B, *et al.*, Anti-neoplastic agents from plants. *Ind J Cancer Chemother* 1985; 7: 59-65.
33. Aschar KRS, Schmulterer H, Glotter E, *et al.*, Distribution of the chemotypes of *Withania somnifiera* in some areas of Israel. Feeding studies with *Spodoptera littoralis* larvae and chemical examination of Withanolide content. *Phytoparasitica* 1984; 12: 147-155.
34. Ghosal S, Lal J, Srivastava R, *et al.*, Immunomodulatory and CNS effects of sitoindosides IX and X, two new glycowithanolides from *Withania somnifera Phytother Res* 1989; 3: 201.
35. Khan FZ, Saeed MA, Alam M, *et al.*, Biological studies of indigenous medicinal plants 111 Phytochemical and antimicrobial studies on the non alkaloidal constituents of some solanaceous fruit. *Fac Pharmacy Gazi University* 1993; 10: 105-116.
36. Kundu AB, Mukerjee A, Dey AK. A new Withanolide from the seeds of *Withania somnifera. Ind J Chem* 1976; 14B: 434-435.
37. Bonetto GM, Gil RR, Oberti JC, *et al.*, Novel Withanolides from *Jaborosa sativa. J Nat Prod* 1995; 58: 705-711.
38. Devi PU, Akagi K, Ostapenko V, *et al.*, Withaferin A: a new radiosensitizer from the Indian medicinal plant *Withania somnifera. Int J Rad Biol* 1996; 69: 193-197.
39. Devi PU. *Withania somnifera* Dunal (Ashwagandha): Potential plant source of a promising drug for cancer chemotherapy and radiosensitization. *Ind J Exp Biol* 1996; 34: 927-932.
40. Sohat B, Gitter E, Abraham A, *et al.*, Antitumor activity of withaferin A. *Can Chemother Rep* 1967; 51: 51271-51276..
41. Kennelly EJ, Gerhäuser C, Song LL, *et al.*, Induction of quinine reductase by Whithanolides isolated from *Physalis philadelphica* (Tomatillos). *Food Chem* 1997; 45: 3771-3777.
42. Kuboyama T, Tohda C, Zhao J, *et al.*, Axon- or dendrite-predominant outgrowth induced by constituents from Ashwagandha. *Neuroreport* 2002; 13: 1715-1720.
43. Mimura H, Gato K, Kitamura S, *et al.*, Effect of Water Content on the Solid-State Stability in two Isomorphic Clathrates of Cephalosporin: Cefazolin Sodium Pentahydrate (α Form) and FK041 Hydrate. *Chem Pharm Bull* 2002; 50: 760-765.

44. Ma L, Xie CM, Li J, *et al.*, Antiproliferative withanolides from MeOH extract of the aerial parts of *Datura metel. Chem Biodiv* 2006; 3: 180-186.
45. Hsieh PW, Huang ZY, Chen JH, *et al.*, Cytotoxic Withanolides from *Tubocapsicum anomalum. J Nat Prod* 2007; 70: 747-753.
46. Lee SW, Pan MH, Chen CM, *et al.*, Withangulatin I: a New Cytotoxic Withanolide from *Physalis angulata. Chem Pharm Bull* 2008; 56: 234-236.
47. He QP, Ma L, Luo JY, *et al.*, Cytotoxic Withanolides from *Physalis angulata* L. *Chem Biodivers* 2007; 4: 443-449.
48. Abdeliebbar LH, Benjouad A, Morjani HH, *et al.*, Antiproliferative Effects of Withanolides from *Withania adpressa. Therapie* 2009; 64: 121-127.
49. Nashi W, Shah N, Didik P, *et al.* Deceleration of Senescence in Normal Human Fibroblasts by Withanone Extracted from Ashwagandha Leaves. J Gerontol Series A: *Biol Sci Med Sci* 2009; 64A: 1031-1038.
50. Cordero CP, Morantes SJ, Páez A, *et al.*, Cytotoxicity of Withanolides isolated from *Acnistus arborescens. Fitoterapia* 2009; 80: 364-368.
51. Pan MR, Chang HC, Wu YC, *et al.*, Tubocapsanolide-A Inhibits Transforming Growth Factor-β-activating Kinase-1 to Suppress NF-κB-induced CCR7. *J Biol Chem* 2009; 284: 2746-2754.
52. Oh JH, Jin LT, Park JW, *et al.*, Withaferin A inhibits INOS expression and nitric oxide production by Akt inactivation and down-regulating LPS-induced activity of NF-kB in RAW 264.7 cells. *Eur J Pharmacol* 2008; 599: 11-17.
53. Jayaprakasam G, Nair B. Cyclooxygenase-2 enzyme inhibitory Withanolides from *Withania somnifera* leaves. *Tetrahedron* 2003; 59: 841-849.
54. Vanisree V, Subbaraju GV, Rao CV, *et al.*, Withanolide sulfoxide from Aswagandha roots inhibits nuclear transcription factor kappa-B, cyclooxygenase and tumor cell proliferation. *Phytother Res* 2009; 23: 987-992.
55. Tohda C, Kuboyama T, Komatsu K. Search for Natural Products Related to Regeneration of the Neuronal Network. *Neurosignals* 2005; 14: 34-45.
56. Tohda KC, Komatsu K. Neuritic regeneration and synaptic reconstruction induced by Withanolide A. *Br J Pharmacol* 2005; 144: 961-971.
57. Mohan R, Hammers HJ, Mohan BP, *et al.*, Withaferin A is a potent inhibitor of angiogenesis. *Angiogenesis* 2004; 7: 115-122.
58. Benjumea D, Herrera DM, Abdala S, *et al.*, Withanolides from *Withania aristata* and their diuretic activity. *J Ethnopharmacol* 2009; 123: 351-355.
59. Maurya R, Jayendra A, Singh AB, *et al.*, Coagulanolide, a Withanolide from *Withania coagulans* fruits and antihyperglycemic activity. *Bioorg Med Chem Lett* 2008; 18: 6534-6537.
60. Huang CF, Ma L, Sun LJ, *et al.*, Immunosuppressive Withanolides from *Withania coagulans. Chem Biodiv* 2009; 6: 1415-1426.
61. Ahmad S, Malik A, Yasmin A, *et al.*, Withanolides form *Physalis peruviana. Phytochemistry* 1999; 50: 647-651.

62. Khan PM, Nawaz HR, Ahmad S, *et al.*, Ajugins C, and D, New Withanolides from *Ajuga parviflora. Helv Chim Acta* 1999; 82: 1423-1426.
63. Nicotra VE, Gil RR, Juan CO, *et al.*, New Withanolides from two varieties of *Jaborosa Caulescens. Molecules* 2000; 5: 514-515.
64. Kinghorn AD, Su BN, Lee D, *et al.*, Cancer Chemopreventive agents discovered by Activity-Guided Fractionation: An Update. *Curr Org Chem* 2003; 7: 213-226.
65. Su BN, Park EJ, Nikolic D, *et al.*, Activity-Guided Isolation of Novel Norwithanolides from *Solanum altissimum* with potential cancer Chemopreventive activity. *J Org Chem* 2003; 68: 2350-2361.
66. Rahman AU, Shahwar ADE, Naz A, *et al.*, Withanolides from *Withania coagulans. Phytochemistry* 2003; 63: 387-390.
67. Shingu K, Furusawa Y, Nohara T. New Withanolides, Daturametelins C, D, E, F and G-Ac from *Datura metel* L. (Solanaceous Studies. XIV). *Chem Pharm Bull* 2005; 37: 2132-2135.
68. Niero R, DaSilva IT, Tonial GC, *et al.*, Cilistepoxide and Cilistadiol, two new Withanolides from *Solanum sisymbiifolium. Nat Prod Res* 2006; 20: 1164–1168.
69. Castorena ALP, Oropeza RF, Vazquez AR, *et al.*, Labdanes and Withanolides from *Physalis coztomatl. J Nat Prod* 2006; 69: 1029-1033.
70. Gu JQ, Li W, Kang YH, *et al.*, Minor Withanolides from *Physalis philadelphica*: Structures, Quinone Reductase Induction Activities, and Liquid Chromatography (LC)-MS-MS Investigation as Artifacts. *Chem Pharm Bull* 2003; 51: 530.
71. Siddiqui BS, Shamsul A, Farhana A, *et al.*, Withanolides from *Datura Innoxia. Heterocycles* 2005; 65: 857-863.
72. Lal P, Mishra L, Sangwan RS, *et al.*, New Withanolides from fresh berries of *Withania somnifera.* Z. *Naturforsch* 2006; 61b: 1143-1147.
73. Subaraju GV, Vanisree M, Rao CV, *et al.*, Ashwagandhanolide, a bioactive dimeric thiowithanolide isolated from the roots of *Withania somnifera. J Nat Prod* 2006; 69: 1790-1792.
74. Pramanick S, Roy A, Ghosh S, *et al.*, Withanolide Z: a new chlorinated Withanolide from *Withania somnifera. Planta Med* 2008; 74: 1745-1748.
75. Ahuja A, Kaur D, Sharada M, *et al.*, Glycowithanolides accumulation in *in vitro* shoot cultures of Indian ginseng (*Withania somnifera* Dunal). *Nat Prod Comm* 2009; 4: 479-482.

Chapter 2

An Insight into Anticancer Mechanism of Withaferin A

2.1 INTRODUCTION

Withania somnifera Dunal, (Solanaceae) usually known as *ashwagandha*, has been used for centuries in Ayurvedic medicine to increase longevity and vitality. Western research supports its polypharmaceutical use, confirming its antioxidant, anti-inflammatory, immunomodulating, and anti-stress properties in the whole plant extract and several separate constituents.[1]

Withanolides are a group of naturally occurring oxygenated ergostane type steroids containing lactone in the side chain and 2-en-1-one system in ring A. Withaferin A (see Chapter 1, Fig. 1.3) is a cell-permeable steroidal lactone from a medicinal plant *W. somnifera*, a plant known in traditional Indian medicine.

Withaferin A is an important withanolide holding promise in cancer treatment and as a relatively safe radiosensitive/chemotherapeutic agent. It is present in traces in all parts of *W. somnifera* except the leaves, where it is reported to be present in only two non-Indian chemo types; AGB002 and AGB025.[2] The productivity of withaferin A in the three week's old cultured roots was reported to be 11.65 μg^{-1}.[3]

2.2 STUDIES ON ANTICANCER ACTIVITY

Withania somnifera—Several reviews highlighting the use of *W. somnifera* and its active constituents as antitumor agents and in conjunction with radiation and chemotherapy treatment have been published. Reversal of paclitaxel induced neutropenia by *W. somnifera* has been reported in mice.[4] Chemotherapeutic efficacy of paclitaxel was enhanced by *W. somnifera* on benzo (a) pyrene-induced experimental lung cancer.[5,6]

An investigatory study reported protective effect of *W. somnifera* extract in Dalton's ascitic lymphoma.[7] The roots of *W. somnifera* have cell cycle disruption and anti-angiogenic activity, which is supposed to be a critical mediator for its anticancer action.[8] A tumor-inhibitory factor with selective killing of cancer cells by the leaf extract of *W. somnifera* has been characterized.[9]

2.3 EXPERIENTIAL STUDIES ON ANTICANCER ACTIVITY OF WITHAFERIN A

2.3.1 Effect of Withaferin A on 180 Tumor Cells

Mouse sarcoma 180 (S-180) solid and ascites tumor cells were treated *in vivo* and *in vitro* with withaferin A. It was found to affect the spindle microtubules of cells in metaphase. An interesting finding was the double membranes surrounding the chromosomes in the treated cells; probably the nuclei were reconstructed directly from the metaphase stage in the *in vivo* withaferin A-treated cells. In addition the membranes of the cells in interphase were affected by *in vivo* or *in vitro* treatment with withaferin A.[10]

2.3.2 Cytotoxic Effects of Withaferin A on P388 Cells

P388 cells, 4-dehydrowithaferin A and withaferin A diacetate exhibited an equal inhibitory effect on thymidine, uridine and L-valine incorporation. They stopped cell proliferation and, at the same time, killed the cells. Withaferin A promptly reacted with L-cysteine, and it was presumed that possible target sites in the cell might be the SH groups of enzymes which react with the lactone and epoxide groups of the agent.[11]

2.3.3 Ehrlich Ascites Carcinoma and Withaferin A

Twenty-four hours after IP inoculation of 10(6) tumor cells, withaferin A was injected IP at different dose fractions (5 or 7.5 mg/kg × 8, 10 mg/kg × 5, 20 or 30 mg/kg × 2) with or without abdominal gamma irradiation (RT, 75. Gy) after the first drug dose. Increase in life span and tumor free survival was studied up to 120 days. Withaferin A inhibited tumor growth and increased survival, which was dependent on the withaferin A dose per fraction rather than the total dose. A combination of RT with all the drug schedules increased tumor cure and tumor-free survival, the best effect were seen after 2 fractions of 30 mg/kg each.[12]

2.3.4 Radiosensitizer Effect in V79 Cells

In a study, withaferin A reduced survival of V79 cells in a dose-dependent manner. LD50 for survival was 16 μM. One-hour treatment with a non-toxic dose of 2.1 μM before irradiation significantly enhanced cell killing, giving a sensitizer enhancement ratio of 1.5 for 37 percent survival and 1.4 for 10 percent survival. The drug induced a G2-M block, with a maximum accumulation of cells in G2-M phase at 4 hr after treatment with 10.5 μM withaferin A in 1 hr.[13]

2.3.5 Role of Withaferin A in Fibrosarcoma and Melanoma

2.3.5.1 Fibrosarcoma and melanoma

Two mouse tumors, B16F1 melanoma and fibrosarcoma were exposed locally to 30 or 50 Gy gamma radiation as an acute dose, or 5 fractions of 10 Gy. Withaferin A, 40 mg/kg, was injected intraperitoneally, 1 hr before acute irradiation, or 30 mg/kg before every 10 Gy fraction. Trimodality treatment synergistically increased complete response to 37 percent in melanoma and to 64 percent in fibrosarcoma. Fractionated radiotherapy (10 Gy × 5) was more effective (25 percent complete response) than acute dose of 50 Gy (0 percent complete response) on melanoma, while there was no difference between the response of fibrosarcoma in the two regimens.[14]

2.3.5.2 Melanoma

In the present investigation, the effect of withaferin A on the development and decay of thermo tolerance in B16F1 melanoma was studied in C57BL mice. Tumors of 10010 mm^3 size were subjected to repeated hyperthermia at 43°C for 30 min. Withaferin A was injected after the first hyperthermia treatment. Tumor growth delay heightened with increase in the time gap between two hyperthermia treatments and was significantly higher ($P < 0.05$ to $P < 0.001$) in withaferin A treated groups.[15]

2.3.6 Inhibition of Angiogenesis

It was proposed that the inhibitory action of withaferin A occurs by interference with the ubiquitin-mediated proteasome pathway as suggested by the increased levels of poly-ubiquitinated proteins. Finally, withaferin A was shown to exert potent anti-angiogenic activity *in vivo.* [16]

2.3.7 Mediation of Action through by Annexin II

The study reported that withaferin A alters cytoskeletal architecture by covalently binding annexin II and stimulating its basal F-actin cross-linking activity. Drug-mediated disruption of F-actin organization is dependent on annexin II expression by cells and markedly limits their migratory and invasive capabilities at subcytotoxic concentrations.[17]

2.3.8 Withaferin A as Proteasome Inhibitor

Withaferin A potently inhibits the chymotrypsin-like activity of a purified rabbit 20S proteasome (IC50 = 4.5 μM) and 26S proteasome in human prostate cancer cultures (at 5-10 μM) and xenografts (4-8 mg/kg/day). Treatment of human prostate PC-3 xenografts with WA for 24 days resulted in 70 percent inhibition of tumor growth in nude mice, associated with 56 percent inhibition of the tumor tissue proteasomal chymotrypsin like activity.[18]

2.3.9 Withaferin A as Apoptosis Inducer

Withaferin A induced Par-4-dependent apoptosis in androgen-refractory prostate cancer cells and regression of PC-3 xenografts in nude mice. Interestingly, restoration of wild-type AR in PC-3 (AR negative) cells abrogated both Par-4 induction and apoptosis by withaferin A. The withanolide and anti-androgen synergistically induced Par-4 and apoptosis in androgen-responsive prostate cancer cells.[19]

2.3.10 Inhibition of Protein Kinase C

In *Leishmania donovani*, the inhibition of protein kinase C by withaferin A causes depolarization of DeltaPsim and generates ROS inside cells. Loss of DeltaPsim leads to the release of cytochrome c into the cytosol and subsequently activates caspase-like proteases and oligonucleosomal DNA cleavage.[20]

2.3.11 Targeting of Protein Vimentin by Withaferin

According to one study, withaferin A binds to the intermediate filament protein, vimentin, by covalently modifying its cysteine residue. Withaferin A induces vimentin filaments to aggregate *in vitro*, an activity manifested *in vivo* as punctate cytoplasmic aggregates that colocalize vimentin and F-actin. Withaferin-A exerts potent dominant-negative effect on F-actin that requires vimentin expression and induces apoptosis.[21]

2.3.12 Withaferin A and Leukemia

2.3.12.1 Human myeloid leukemia and withaferin A

Withaferin A primarily induces oxidative stress in human leukemia HL-60 cells and in several other cancer cell lines. The withanolide induces early ROS generation and mitochondrial membrane potential (Deltapsi(mt)) loss, which precedes release of cytochrome c, translocation of Bax to mitochondria and apoptosis inducing factor to cell nuclei. These events paralleled activation of caspases-9, -3 and PARP cleavage.[22]

2.3.12.2 Leukemia

Withaferin A induces apoptosis in association with the activation of caspase-3. JNK and Akt signal pathways play crucial roles in withaferin A-induced apoptosis in U937 cells. Over expression of Bcl-2 and active Akt (myr-Akt) in U937 cells inhibited the induction of apoptosis, activation of caspase-3, and PLC-gamma 1 cleavage by withaferin A.[23]

2.3.13 NFkappaB and Withaferin A

Leaf extract of *W. somnifera* as well as withaferin A, potently inhibits NFκB activation by preventing the tumor necrosis factor-induced activation of IκB kinase beta via a thioalkylation-sensitive redox mechanism. This prevents IκB phosphorylation and degradation, which subsequently blocks NFκB translocation, NFκB/DNA binding, and gene transcription.[24]

2.3.14 Oral Carcinogenesis and Withaferin A

Oral administration of withaferin A (20 mg/kg body weight) to 7,12-dimethylbenz[a]anthracene administered to animals for 14 weeks completely prevented tumor incidence, volume and burden. Also, Withaferin A showed significant anti-lipid peroxidative and anti-oxidant properties and maintained the status of phase-I and phase-II detoxication agents during MBA induced oral carcinogenesis.[25]

2.3.15 Effect in Breast Cancer Cells

Treatment of MDA-MB-231 (estrogen-independent) and MCF-7 (estrogen-responsive) cell lines with withaferin A resulted in a concentration and time-dependent increase in G2-M fraction, which correlated with a decrease in levels of cyclin-dependent kinase 1 (Cdk1), cell division cycle 25C (Cdc25C) and/or Cdc25B proteins, leading to accumulation of Tyrosine 15 phosphorylated (inactive) Cdk1.[26]

2.3.16 Withanolides and Gliomas

Withaferin A, withanone, withanolide A and the leaf extract of *W. somnifera* markedly inhibited the proliferation of glioma cells in a dose dependent manner and changed their morphology toward the astrocytic type. Molecular analysis revealed that the leaf extract of *W. somnifera* and some of its components caused enhanced expression of glial fibrillary acidic protein, change in the immunostaining pattern of mortalin from perinuclear to pancytoplasmic, delay in cell migration and increased expression of neuronal cell adhesion molecules.[27]

2.3.17 Inhibition of Notch-1 by Withaferin A

Withaferin A inhibited Notch-1 signaling and downregulates prosurvival pathways, such as Akt/NF-κB/Bcl-2, in three colon cancer cell lines (HCT-116, SW-480, and SW-620). In addition, it downregulates the expression of mammalian target of rapamycin signaling components, pS6K and p4EBP1, and activates c-Jun-NH (2)-kinase-mediated apoptosis in colon cancer cells.[28]

2.3.18 Pancreatic Cancer

Withaferin A exhibited potent antiproliferative activity against pancreatic cancer cells *in vitro* (with IC50 s of 1.24, 2.93 and 2.78 µM) in pancreatic cancer cell lines Panc-1, MiaPaCa2 and BxPc3, respectively. Withaferin A —biotin binds to C-terminus of Hsp90 which is competitively blocked by unlabeled withaferin A. Withaferin A—(3, 6 mg/kg), inhibited tumor growth in pancreatic Panc-1 xenografts by 30 and 58 percent, respectively.[29]

2.3.19 Neck Squamous Carcinoma

A study showed that withaferin A extracted from the aerial parts of *Vassobia breviflora* (Sendtn.) Hunz. (Solanaceae) induces apoptosis and cell death in neck squamous cell carcinoma cells as well as a cell-cycle shift from G0/G1 to G2-M. Cells treated with withaferin A exhibited inactivation of Akt and a reduction in total Akt concentration.[30]

REFERENCES

1. Singh AP. Withaferin A: a Potential Anticancer Withanolide from *Withania somnifera* (L.) Dun.: *Ethnobot Leaflets* 2004; 1: 888-889.
2. Kaul MK, Kumar A, Mir BA, *et al.* Production dynamics of Withaferin A in *Withania somnifera* (L.) Dunal complex. *Nat Prod Res* 2009; 23: 1304-1311.
3. AbouZid SF, El-Bassuony AA, Nasib A, *et al.* Withaferin A: Production by Root Cultures of *Withania coagulans. Int J Appl Res Nat Prod* 2010; 3: 23-27.
4. Gupta YK, Sharma SS, Rai K, *et al.* Reversal of paclitaxel induced neutropenia by *Withania somnifera* in mice. *Indian J Physiol Pharmacol* 2001; 45: 253-257.
5. Senthinathan P, Padmavati R, Magesh V, *et al.* Chemotherapeutic efficacy of paclitaxel in combination with *Withania somnifera* on benzo(a)pyrene-induced experimental lung cancer. *Cancer Sci* 2006; 97: 654-658.
6. Senthinathan P, Padmavati R, Banu SM, *et al.* Enhancement of antitumor effect of paclitaxel in combination with immunomodulatory *Withania*

somnifera on benzo(a)pyrene induced experimental lung cancer. *Chem Biol Interact* 2006; 59: 180-185.
7. Christina AJ, Joseph DG, Packialakshmi M, *et al.* Anticarcinogenic activity of *Withania somnifera* Dunal against Dalton's Ascitic Lymphoma. *J Ethnopharmacol* 2004; 93: 359-361.
8. Mathur R, Gupta SK, Singh N, *et al*. Evaluation of the effect of *Withania somnifera* root extracts on cell cycle and angiogenesis. *J Ethnopharmacol* 2006; 105: 336-341.
9. Widodo N, Kaur K, Shrestha BG, *et al*. Selective killing of cancer cells by leaf extract of Ashwagandha: identification of a tumorinhibitory factor and the first molecular insights to its effect. *Clin Cancer Res* 2007; 13: 2298-2306.
10. Shohat B, Shaltiel A, Ben-Bassat M, *et al.* The effect of withaferin A on the fine structure of S-180 tumor cells. *Cancer Lett* 1976; 2: 71-77.
11. Fuskova A, Fuska J, Rosazza JP, *et al.* Novel cytotoxic and antitumor agents. IV. Withaferin A: relation of its structure to the *in vitro* cytotoxic effects on P388 cells: *Neoplasma* 1984; 31: 31-36.
12. Sharada AC, Solomon FE, Devi PU, *et al.* Antitumor and radiosensitizing effects of withaferin A on mouse Ehrlich ascites carcinoma *in vivo. Acta Oncol* 1996; 35: 95-100.
13. Devi PU, Akagi K, Ostapenko V, *et al.* Withaferin A: a new radiosensitizer from the Indian medicinal plant *Withania somnifera. Int J Rad Biol* 1996; 69: 193-197.
14. Devi PU, Kamath R. Radiosensitizing effect of withaferin A combined with hyperthermia on mouse fibrosarcoma and melanoma. *J Rad Res* 2003; 44: 1-6.
15. Kalthur G, Mutalik S, Pathirissery UD. Development and Decay of Thermo tolerance in B16F1 Melanoma: A Preliminary Study. *Integ Cancer Ther* 2009; 8: 93-97.
16. Mohan R, Hammers HJ, Bargagna-Mohan P, *et al.* Withaferin A is a potent inhibitor of angiogenesis. *Angiogenesis* 2004; 7: 115-122.
17. Falsey RR, Marron MT, Gunahearth GM, *et al.* Actin microfilament aggregation induced by withaferin A is mediated by annexin II. *Nat Chem Biol* 2006; 2: 33-38.
18. Yang H, Shi G, Dou QP. The tumor proteasome is a primary target for the natural anticancer compound Withaferin A isolated from "Indian winter cherry". *Mol Pharmacol* 2007; 71: 426-437.
19. Srinivasan S, Ranga RS, Burikhanov R, *et al.* Par-4-dependent apoptosis by withaferin A in prostate cancer cells. *Cancer Res* 2007, 67: 246-253.
20. Sen N, Banerjee B, Das BB, *et al.* Apoptosis is induced in leishmanial cells by a novel protein kinaseinhibitor withaferin A and is facilitated by apoptotic topoisomerase I-DNA complex. *Cell Death Differ* 2007; 14: 358-367.

21. Bargagna-Mohan P, Hamza A, Kim YE, *et al.* The tumor inhibitor and antiangiogenic agent withaferin A targets the intermediate filament protein vimentin. *Chem Biol* 2007; 14: 623-634.
22. Malik F, Kumar A, Bhushan S, *et al.* Reactive oxygen species generation and mitochondrial dysfunction in the apoptotic cell death of human myeloid leukemia HL-60 cells by withaferin A with concomitant protection by N-acetyl cysteine. *Apoptosis* 2007; 12: 2115-2133.
23. Oh JH, Lee TJ, Kim SH, *et al.* Induction of apoptosis by withaferin A in human leukemia U937 cells through downregulation of Akt phosphorylation. *Apoptosis* 2008; 13: 1494-1504.
24. Kaileh M, Vanden Berghe W, Heyerick A, *et al.* Withaferin a strongly elicits IκB kinase beta hyperphosphorylation concomitant with potent inhibition of its kinase activity. *J Biol Chem* 2007; 282: 4253-4257.
25. Manoharan S, Panjamurthy K, Menon VP, *et al.* Protective effect of Withaferin A on tumour formation in 7, 12-dimethylbenz[a]anthracene induced oral carcinogenesis in hamsters. *Indian J Exper Biol* 2009; 47: 16-23.
26. Stan SD, Zeng Y, Singh SV. Ayurvedic medicine constituent withaferin a causes G2 and M phase cell cycle arrest in human breast cancer cells. *Nutr Cancer* 2008; 60: 51-60.
27. Shah N, Kataria H, Kaul SC, *et al.* Effect of the alcoholic extract of Ashwagandha leaves and its components on proliferation, migration, and differentiation of glioblastoma cells: combinational approach for enhanced differentiation. *Cancer Sci* 2009; 100: 1740-1747.
28. Koduru S, Kumar R, Srinivasan S, *et al.* Notch-1 inhibition by Withaferin A: a therapeutic target against colon carcinogenesis. *Mol Cancer Ther* 2010; 9: 202-210.
29. Yu Y, Hamza T, Zhang T, *et al.* Withaferin A targets heat shock protein 90 in pancreatic cancer cells. *Biochem Pharmacol* 2010; 79: 542-551.
30. Samadi AK, Tong X, Mukerji R, *et al.* Withaferin A, a Cytotoxic Steroid from *Vassobia breviflora*, Induces Apoptosis in Human Head and Neck Squamous Cell Carcinoma. *J Nat Prod* 2010; 73: 1476-1481.

Chapter 3

Complementary and Alternative Medicine Approaches in the Treatment of Cancer

3.1 INTRODUCTION

Alternative medicine is often used instead of standard medical treatment; sometimes with serious consequences for the patient.[1] Complementary medicine is used alongwith with mainstream medical care. Some complementary methods can also cause harm, but if chosen carefully and used properly, they might improve the quality of life of the patient.[2]

According to a comprehensive survey of Americans using CAM, 36 percent of US adults were found to be using some form of CAM.[3] The survey found that rates of CAM use are especially high among patients with serious illnesses such as cancer. When megavitamin therapy, and prayer for health reasons are included in the definition of CAM this percentage rises to 62 percent. Sixty nine percent of 453 cancer patients have used at least one CAM therapy as part of their cancer treatment. Eighty eight percent of 102 people with cancer have used one of the CAM therapies. Of these 93 percent used supplements, 53 percent used non supplement forms of CAM and 47 percent have used both.[4]

3.2 COMPLEMENTARY AND ALTERNATIVE METHODS FOR CANCER MANAGEMENT[5]

People with cancer sometimes consider CAM for a number of reasons:

1. To relieve the side-effects of main stream cancer treatment.
2. To find a less unpleasant approach that has few or no side effects.
3. To take an active role in improving their own health and wellbeing.

Some complementary therapies may help to relieve certain symptoms of cancer, relieve side-effects of cancer therapy, or improve a patient's sense of well being. The American Cancer Society urges patients who are considering the use of any alternative or complementary therapy to discuss this with their health care team. Some alternative therapies have dangerous or even life-threatening side-effects. With others, the main danger being that the patient may be unable to benefit from standard therapy.[5]

Complementary approaches that are used with cancer treatment include aromatherapy, biofeedback, botanical or herbal medicines, cartilage therapy, labyrinth walking, music therapy, Garson therapy, Essiac herbal formula and nutritional supplements. Among several CAM anticancer approaches, botanical or herbal medicines are widely used. There is some evidence of botanical or herbal medicines and their role in cancer treatment. The role of the majority of CAM approaches, in cancer treatment is not scientifically validated.[6]

3.3 ESSIAC HERBAL FORMULA[7,8]

The formula came from the Ojibwe people of North America, one of whom gave it to a white woman around the end of the last century. She in turn gave it to a highly qualified and experienced nurse in an Ontario hospital in 1992. The formula has been reported to be an effective immunomodulator. Essiac does three things—It cleanses the blood, cleans the liver and oxygenates the cells. As a result, a sick person feels more energetic and euphoric Ingredients of Essiac herbal formula (Table 3.1 and Fig. 3.1) are as follows:

- Burdock, root (*Arctium lappa*)
- Sheep sorrel, leaf (*Rumex acetosella*)
- Slippery Elm, powder (*Ulmus fulva*)
- Turkey Rhubarb, root powder (*Rheum officinale*)

Table 3.1 Phytochemical profile of Essiac herbal formula

S. no	*Medicinal Plant*	*Part used*	*Phytochemistry*
1.	*Arctium lappa* L. (Asteraceae)	Roots	Bitter glycoside: lappatin, sesquiterpene lactone: arctopicrin, fukinone, lignans: arctin, arctigenin, mateiresionol, arctiol, lappol, sterol: taraxasterol
2.	*Rheum officinale* L. (Polygonaceae)	Roots	Antharquinone derivatives: chrysophanic acid, emodin, rhein, calcium oxalate, oxalic acid and resinous substance
3.	*Rumex acetosella* L. (Chenopodiaceae)	Roots	Antharquinone derivatives: lapathin and lapathin, calcium oxalate, oxalic acid, and resinous substance
4.	*Ulmus fulva* L. (Ulmaceae)		Tannins

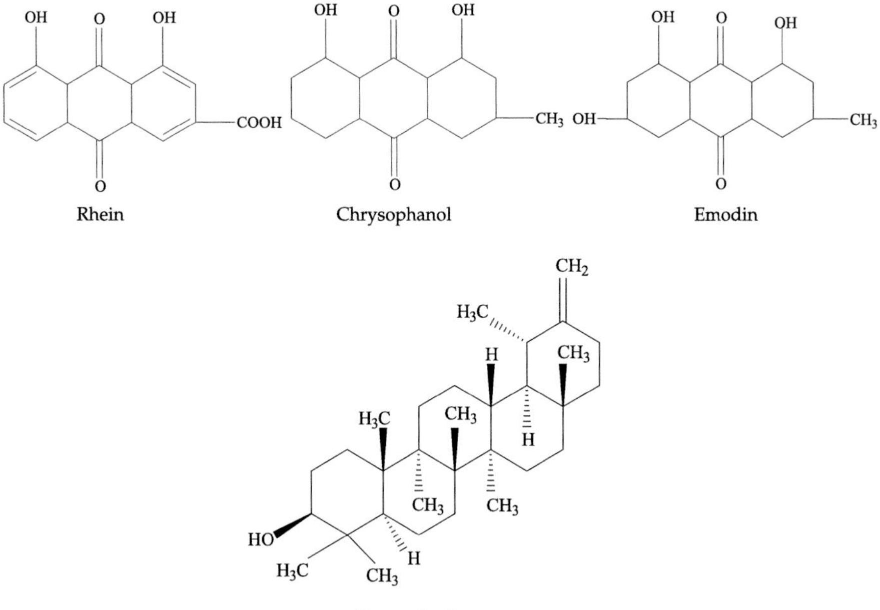

Rhein Chrysophanol Emodin

Taraxasterol

Figure 3.1 Major phytochemicals of Essiac herbal formula.

3.4 CARTILAGE

A cartilage is a type of tough, flexible connective tissue that forms parts of the skeleton in many animals. The cartilage contains cells called chondrocytes, which are surrounded by collagen and proteoglycans, which are made of protein and carbohydrates. Three theories have been proposed to explain how the cartilage acts against cancer

- As the cartilage is broken down by the body, it releases products* that kill cancer cells.
- The cartilage increases the action of the body's immune system to kill cancer cells.
- The cartilage makes substances that block tumor angiogenesis (the growth of new blood vessels that feed a tumor and help it grow.

Based on laboratory and animals tests, the third theory is the most likely. The cartilage does not contain blood vessels, so cancer cannot easily grow in it. It is suggested that a cancer treatment using the cartilage may keep blood vessels from forming in a tumor, causing the tumor to stop growing or shrinking.

In animal studies, cartilage products have been given by mouth; injected into a vein or the abdomen; applied to the skin; or placed in slow-release plastic pellets that were surgically implanted. In studies with humans; cartilage products have been given by mouth; applied to the skin;

*Proteins or glycoproteins or glycosaminoglycans

injected under the skin; or given by enema. The dose and length of time of the cartilage treatment was different for each study, in part because of different types of products that were used.

3.5 GERSON THERAPY[9]

Gerson therapy has been used by some people to treat cancer and other diseases. It is based on the role of minerals, enzymes and other dietary factors. There are three key parts to the therapy:

- **Diet:** Organic fruits, vegetables and whole grains to give the body plenty of vitamins, minerals, enzymes and other nutrients. Fruits and vegetables are low in sodium and high in potassium.
- **Supplementation:** The addition of certain substances in the diet to help correct cell metabolism (the chemical changes that take place in a cell to make energy and basic materials needed for the body's life processes).
- **Detoxification:** Treatments, including enemas, to remove toxic substances from the body.

The Gerson therapy is based on the idea that cancer develops when there are changes in cell metabolism because of the buildup of toxic substances in the body. According to Gerson, the disease process makes more toxins and the liver becomes overloaded . According to Dr. Gerson, people with cancer also have too much sodium and too little potassium in the cells of their bodies, which causes tissue damage and weakened organs.

The goal of Gerson's therapy is to restore the body to health by repairing the liver and returning the metabolism to its normal state. It can be done by removing toxins from the body and building up the immune system with diet and supplements. The enemas are said to widen the bile ducts of the hepatobiliary apparatus so that toxins can be released. The liver is further overworked as the treatment regimen breaks down the cancer cells and helps the body to get rid of toxins. Pancreatic enzymes are given to decrease the demands on the weakened liver and pancreas to make enzymes for digestion.

3.6 SPECIFIC BOTANICALS

3.6.1 Mistletoe Extracts

Research study is being conducted to find out whether an extract of the European mistletoe plant (*Viscum album* L.), along with a chemotherapy drug called gemcitabine, can help treat people with certain cancers.[10]

3.6.2 Dietary Phytochemicals

Recently, numerous reviews of plant derived chemopreventive compounds have been identified for their role in the treatment of cancer. These chemopreventive compounds, usually known as phytopharmaceuticals, are dietary ingredients, which are derived from food and considered pharmacologically safe.[11] Some of the common chemopreventive dietary compounds derived from dietary ingredients are shown in Fig. 3.2.

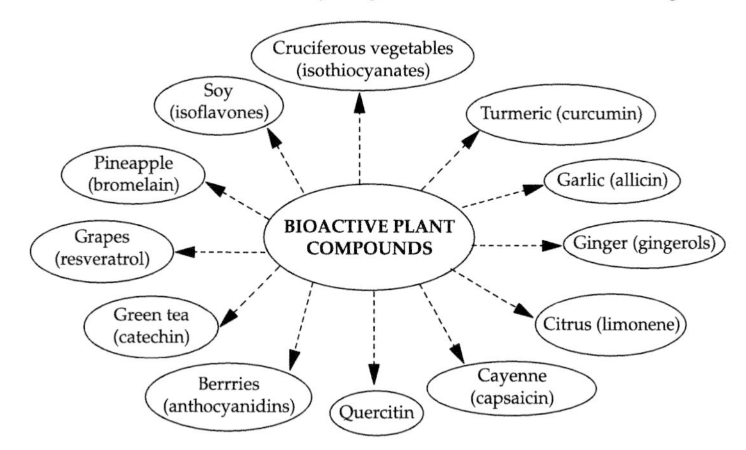

Figure 3.2 Common chemopreventive dietary compounds.

Chemopreventive plant compounds affect all phases of the cancer process, i.e., tumor initiation, promotion and progression. Botanical medicines are complex natural mixtures of pharmacological multitaskers, simultaneously exerting influence on different levels and via different mechanisms. By contrast, pharmaceutical drugs are classically single synthetic compounds, ideally interfering or disrupting a specific mechanism.[12]

3.6.2.1 *β*-sitosterol

β-sitosterol is a phytosterol. Epidemiological and experimental studies have suggested a protective role of *β*-sitosterol (Fig. 3.3) in the development of some types of cancer such as breast, colon and prostate. *In vivo* studies have shown that *β*-sitosterol inhibits proliferation and induces apoptosis in colon and breast cancers. The studies clearly show that *β*-sitosterol kills breast cancer cells and is not toxic to the normal cell. Clinical studies linking *β*-sitosterol and breast cancer are still lacking but some scientists suggest that it may improve the efficiency of tamoxifen, a drug used to treat breast cancer.[13]

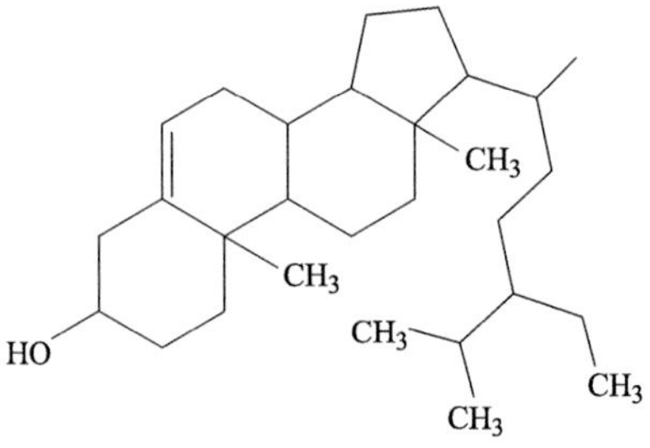

Figure 3.3 Structure of *β*-sitosterol.

3.6.2.2 Curcumin

Dietary administration of curcumin (Fig. 3.4) to mice at a level of 2 percent reduced the incidence of experimentally-induced colonic hyperplasia, indicating that the antioxidant effects are active *in vivo*. Curcumin inhibits cancer at initiation, promotion and progression stages of development.

Research has demonstrated that curcumin blocks the activity of a hormone having a link with the development of colorectal cancer. Other studies have demonstrated that curcumin inhibits melanoma cell growth and destroys cancer cells. In addition, animal research reported that curcumin prevented the spread of breast cancer cells to lungs.[5,16]

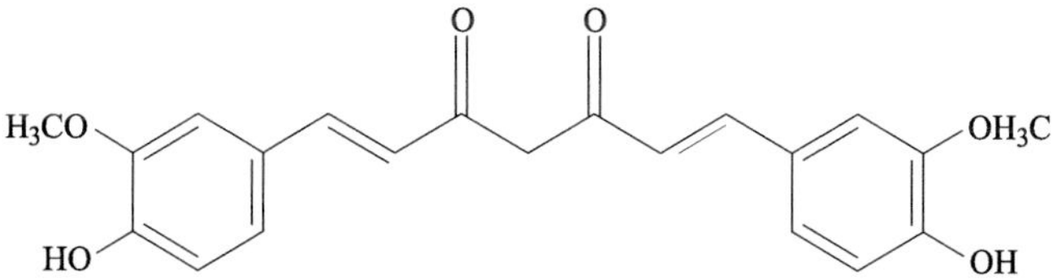

Figure 3.4 Structure of curcumin.

3.6.2.3 (–)-Epigallocatechin gallate

(–)-Epigallocatechin gallate (Fig. 3.5) has an astringent effect and may inhibit cell membrane phosphorylation. However, researchers do not know whether the polyphenols inhibit the initiation or the promotion of tumors. Tea also contains caffeine at a significant level (about 5 percent) and this has been shown to have a small tumor inhibiting effect. While this is not confirmed, but the study recommends a relaxing cup of tea anyway.[14]

In this study the effects of green tea and its major components, (–)-epigallocatechin gallate and caffeine (Fig. 3.6), on the tobacco-specific nitrosamine 4-(methylnitrosamino)-1-(3-pyridyl)-1-butanone (NNK)-induced lung tumorigenesis in A/J mice was examined. Inhibition by green tea and (–)-Epigallocatechin gallate in (NNK)-induced lung tumorigenesis is due at least partly to their antioxidant properties.[15]

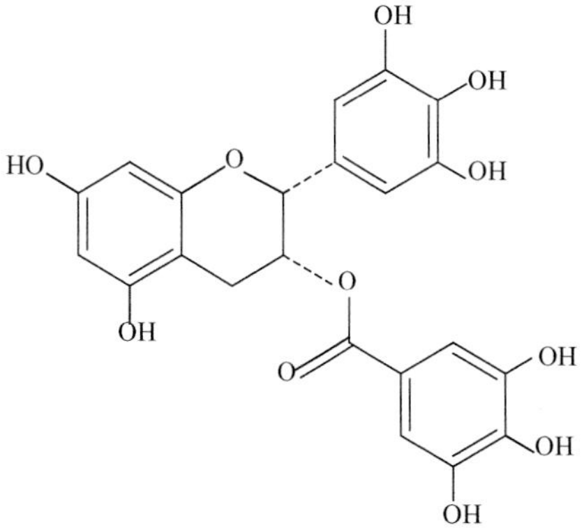

Figure 3.5 Structure of (–)-Epigallocatechin gallate.

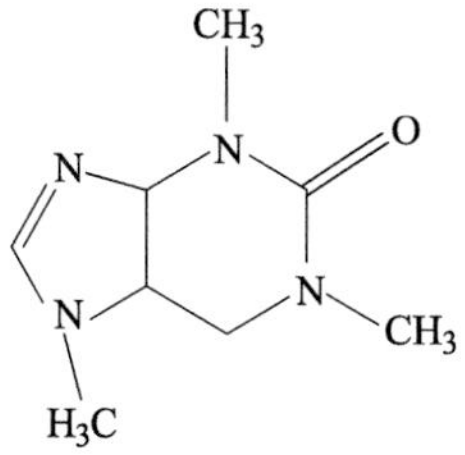

Figure 3.6 Structure of caffeine.

3.6.2.4 Resveratrol

It is a type of polyphenol known as phytoalexin, a class of compounds produced as part of a plant's defense system against disease. Resveratrol (Fig. 3.7) has been shown to reduce tumor incidence in animals by affecting one or more stages of cancer development. It has been shown to inhibit growth of many types of cancer cells in culture. There is evidence that it can reduce inflammation. It also reduces activation of NFκB, a protein produced by the body's immune system when it is under attack. This protein affects cancer cell growth and metastasis.[5,16]

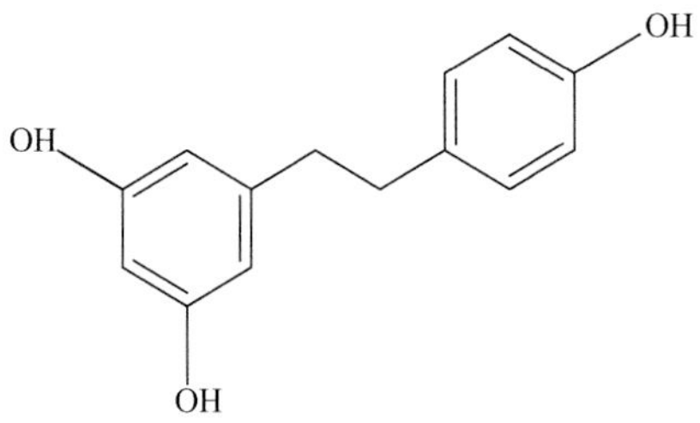

OH
OH
OH

Figure 3.7 Structure of resveratrol.

3.6.2.5 Phytic acid

Phytic acid has a (Fig. 3.8) chelating effect that may serve to prevent, inhibit or even cure some cancers by depriving these cells of minerals, especially iron which is needed for reproduction. The deprivation of essential minerals like iron could, like other broad treatments for cancer, also have negative effects on ono-cancerous cells.[5,16]

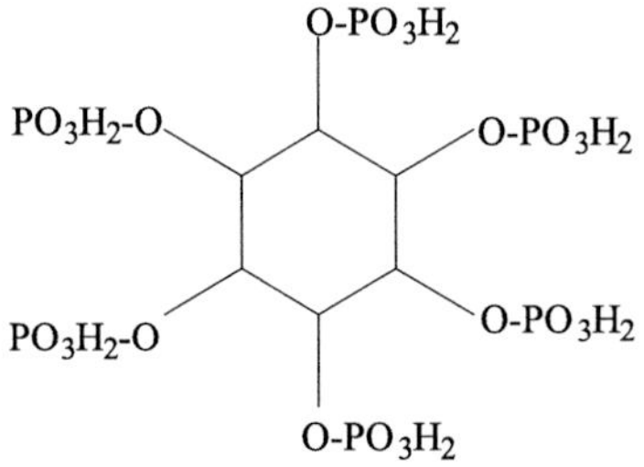

$O\text{-}PO_3H_2$
$PO_3H_2\text{-}O$
$O\text{-}PO_3H_2$
$PO_3H_2\text{-}O$
$O\text{-}PO_3H_2$
$O\text{-}PO_3H_2$

Figure 3.8 Structure of phytic acid.

3.6.2.6 Punicalagin

A recent animal study reported that ellagitannins present in fruit of pomegranates have a possible protective role in prostate cancer. The researchers found that ellagitannins accumulate in the prostate and may be the mode of cancer prevention action. Punicalagin was suggested as possible anticancer agent.[16]

3.6.2.7 S-allyl cysteine

The key ingredient in garlic is S-allyl cysteine, which has been proven to protect against oxidation, free radicals, pollution, cancer and cardiovascular diseases. It was found that S-allyl cysteine derived from Aged Garlic Extract inhibited the proliferation of nine human melanoma cell lines and one murine melanoma cell line in a dose dependent manner. S-allyl cysteine inhibited cellular growth and proliferation and modulated major cell differentiation marker of melanoma.[17]

3.6.2.8 Phytochemicals of *Persea americana* Mill. (Avocado)

Recent studies reported that phytochemicals found in the fruit can prevent the onset of cancer and kill some cancer cells. Phytochemicals extracted from the fruit strike the multiple signaling pathways and prevent cancer by inducing diseases cell death. The phytochemicals has no effect on healthy cells. The fruit contains proteins (25 percent) vitamin C, vitamin E, unsaturated fatty acids and sesquiterpenes. The fruit has no sodium.[18]

3.7 CONCLUSION

Complementary and alternative therapies help to relieve certain symptoms of cancer, reduce the side-effects of cancer therapy, or improve a patient's sense of well being. Side-effects and the ecomoncis of conebtional anticancer drugs have inspired scientists to study plant or natural products or CAM therapies used in cancer, for potential and cost-effective cures. The results obtained from studies with dietary phytochemcials are noteworthy and need further attention.

REFERENCES

1. World Health Organisation. *The promotion and development of traditional medicine.* Geneva: World Health Organization, 1978. (Technical reports series no. 622).
2. Wasik J. *The truth about herbal supplements.* Consumers Digest. July/August 1999: 75-79.
3. Flynn R, Roest M. *Your guide to standardized herbal products.* Prescott, AZ: One World Press, 1995.
4. Aggawal BB, Takada Y, Oommen OV. From chemoprevention to chemotherapy: common targets and common goals. *Exp Opin Invest Drug* 2004, 13: 1327-1338.
5. Bagchi D, Preuss H. *Phytopharmaceuticals in Cancer Chemoprevention.* Boca Raton: CRC Press, 2005.
6. Weiner MA, Weiner JA. *Herbs That Heal.* Quantum Books, Mill Valley, California, 1994.

7. Hoad J. *Healing with Herbs.* Jaico Publishing House, New Dehli, 1997; pp: 84-86.
8. Bell EA, Charlwood BV. *Secondary plant products* (Encyclop. Plant Physiology, Vol. 8) Berlin-Heidelberg-New York: Springer Verlag, 1980.
9. Readers Digest: "The healing power of vitamins, minerals and herbs publishers by Readers digest Australia, Reprinted. 2001-2002.
10. Mansky PJ, Grem J, Wallerstedt DB, Monahan BP, Blackman MR. Mistletoe and gemcitabine in patients with advanced cancer: a model for the phase I study of botanicals and botanical-drug interactions in cancer therapy. *Integr Cancer Ther* 2003; 2: 345-352.
11. Bagchi D, Preuss H. *Phytopharmaceuticals in Cancer Chemoprevention.* Boca Raton: CRC Press; 2005.
12. Anonyms. The role of phytochemicals in optimal Health. *J Nat Acade Child Develop*, 1997; 2.
13. Award AB, Barta SL, Fink CS, *et al.* *β*-sitosterol enhances tamoxifen effectiveness on breast cancer cells by affedting ceramide metabolism. *J Mol Nutr Food Res* 2008; 12: 12-19.
14. Zhao BL, Li XJ, He RG, *et al.* Scavenging effect of extracts of green tea and natural antioxidants on active oxygen radicals. *Cell Biophys* 1989; 14: 175-185.
15. Xu Y, Ho CT, Amin SG, *et al.* Inhibition of tobacco-specific nitrosamine-induced lung tumorigenesis in A/J mice by green tea and its major polyphenol as antioxidants. *Cancer Res* 1992; 52: 3875-3879.
16. Kuo P-Lin, Hsu Ya-Ling, Lin C. The Chemopreventive effects of natural products against human cancer cells. *Int J App Sci Eng* 2005; 3: 203-214.
17. Takeyama H, Hoon DS, Saxton RE, *et al.* Growth Inhibition and Modulation of Cell Markers of Melanoma by S-allyl cysteine. *Oncology* 1993; 50: 63-69.
18. Ding H, Chin YW, Kinghorn AD, D'Ambrosio SM. Chemo preventive characteristics of avocado fruit. *Semin Cancer Biol* 2007; 17(5): 386-394.

Chapter 4

Review of Anticancer and Cytotoxic Potential of Sesquiterpenoids

4.1 IXERIN Z AND 11,13A-DIHYDROIXERIN Z

The antitumor effects of sesquiterpene lactone glycosides ixerin Z and 11,13A-dihydroixerin Z (Fig. 4.1), two known sesquiterpene lactone glycosides isolated from *Crepidiastrum sonchifolium* (Bunge) J.H. Pak & Kawano (Asteraceae), and sesquiterpene lactone aglycones 3-hydroxy-1(10),3,11(13)-guaiatriene-12,6-olide-2-one and 3-hydroxy-1(10),3-guaiadiene-12,6-olide-2-one were measured *in vitro* and *in vivo*.

Sesquiterpene lactone aglycones inhibited various cultured cell growth in a dose-dependent manner as indicated by MTT assay. The possible mechanism of cytotoxicity was investigated on aglycones with the observation that they could affect DNA replication by inhibiting the ssDNA binding activity of replication protein A with the use of simian virus (SV40) DNA *in vitro* replication system. Furthermore, it was also revealed in cell cycle analysis that aglycones could arrest A549 cells in G2-M phase.[1]

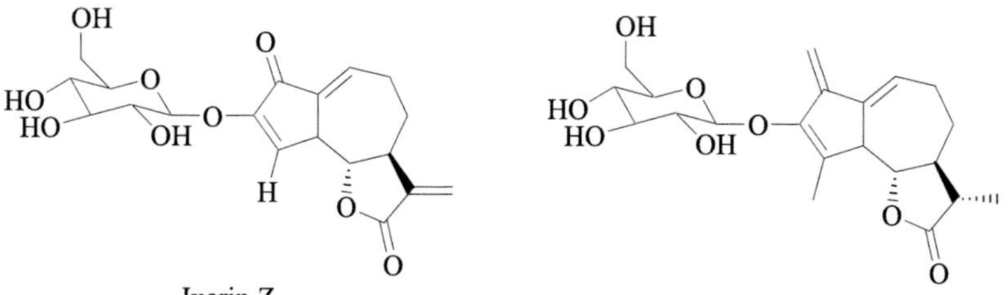

Ixerin Z

11,13a-dihydroixerin Z

Figure 4.1 Sesquiterpene lactone glycosides of *Crepidiastrum sonchifolium*.

4.2 TORILIN, 1β-HYDROXYTORILIN, AND 1α-HYDROXYTORILIN

Guaiane-type sesquiterpenoids, torilin, 1-β-hydroxytorilin, and 1-α-hydroxytorilin (Fig. 4.2) isolated from the methylene chloride-soluble fraction of the methanolic extract of the fruits of *Torilis japonica* (Houtt.) D.C. (Umbelliferae) exhibited cytotoxicity against human A549, SK-OV-3, SK-MEL-2, and HCT15 tumor cells.[2]

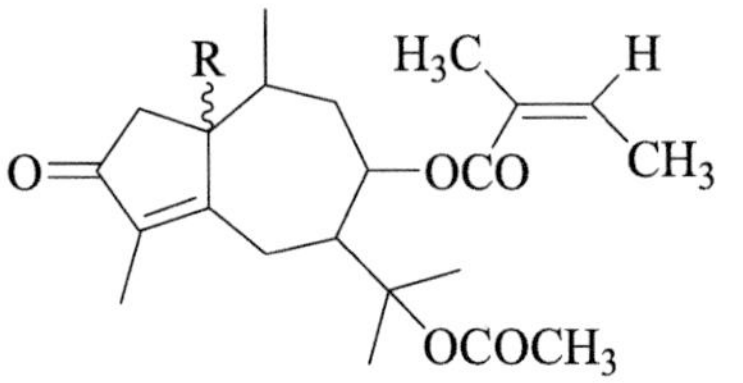

Compounds	*R*
1	β - H; Torilin
2	α - OH; 1-α-hydroxytorilin
3	β - OH; 1-β-hydroxytorilin

Figure 4.2 Guaiane-type sesquiterpenoids of *Torilis japonica.*

4.3 SCABERTOPIN, DEOXYELEPHANTOPIN AND ISODEOXYELEPHANTOPIN

Sesquiterpene lactones, scabertopin, deoxyelephantopin and isodeoxyelephantopin (Fig. 4.3) isolated from *Elephantopus scaber* Linn. (Asteraceae) exhibited significant antitumor effect *in vitro* in a concentration-dependent manner. Deoxyelephantopin also possessed antitumor activity *in vivo*.[3]

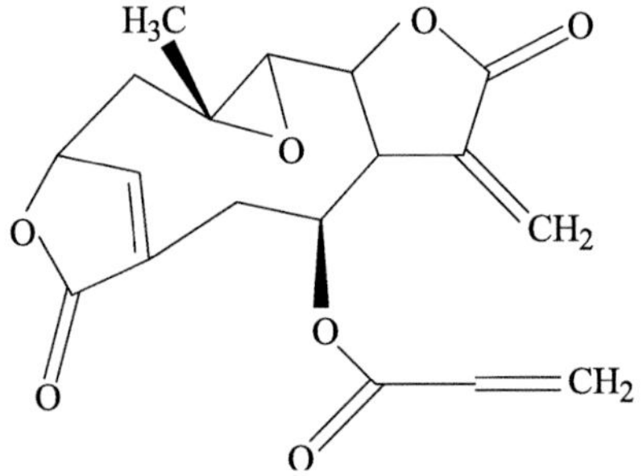

Figure 4.3 Structure of Deoxyelephantopin.

4.4 INULACAPPOLIDE

In vitro, inulacappolide (Fig. 4.4), a new germacranolide isolated from the EtOH extract of the whole plant of *Inula cappa* (Buch.-Ham. ex D. Don) DC. (Asteraceae) showed antiproliferative effects against human cervical cancer HeLa, human leukemia K562 and human nasopharyngeal carcinoma KB cell lines with IC50 values of 1.2 μM, 3.8 μM and 5.3 μM, respectively.[4]

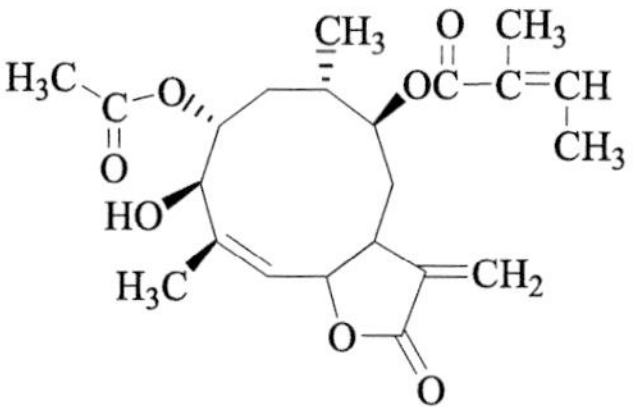

2α-acetoxy-3β-hydroxy-9β-angeloyloxygermacra-4-en-6α, 12-olide

Figure 4.4 Chemical structure of inulacappolide.

4.5 GERMACRANOLIDE SESQUITERPENE LACTONES ISOLATED FROM *CARPESIUM TRISTE* VAR. *MANSHURICUM*

Four known germacranolide sesquiterpene lactones (Fig. 4.5) isolated from *Carpesium triste* var. *manshuricum* (Asteraceae) showed significant cytotoxicities (ED50 value: 4.3-16.8 μM) against five human tumor cell lines; A549, SK-OV-3, SK-MEL-2, XF498 and HCT15.[5]

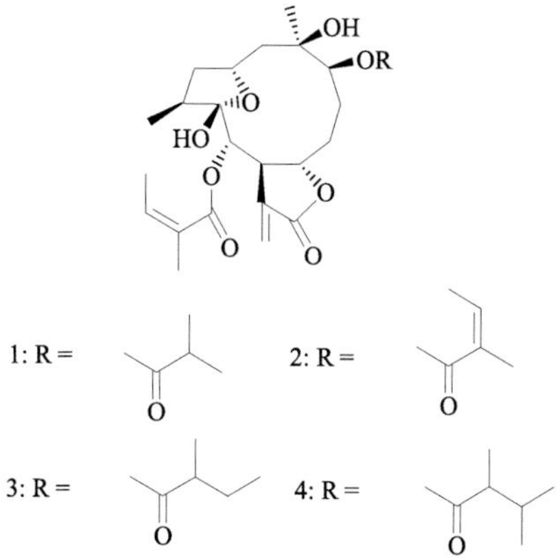

1: 2α,5-Epoxy-5,10-dihydroxy-6α-angeloyloxy-9β-isobutyloxy-germacran-8α,12-olide
2: 2α,5-Epoxy-5,10-dihydroxy-6α,9β-diangeloyloxy-germacran-8α,12-olide
3: 2α,5-Epoxy-5,10-dihydroxy-6α-angeloyloxy-9β-(2-methylbutyloxy)-germacran-8α,12-olide
4: 2α,5-Epoxy-5,10-dihydroxy-6α-angeloyloxy-9β-(3-methylbutyloxy)-germacran-8α,12-olide

Figure 4.5 Germacranolide sesquiterpene lactones isolated from *Carpesium triste* var. *manshuricum* (Asteraceae).

4.6 COSTUNOLIDE, β-CYCLOCOSTUNOLIDE, DIHYDRO COSTUNOLIDE AND DEHYDRO COSTUSLACTONE

Dried roots of *Saussurea lappa* are used in traditional medicine for the treatment of cancer. A new sesquiterpene isolated from *S. lappa* exhibited potent cytotoxic activity. The known compounds costunolide, *β*-cyclocostunolide, dihydro costunolide and dehydro costuslactone (Fig. 4.6) exhibited moderate cytotoxic activity.[6]

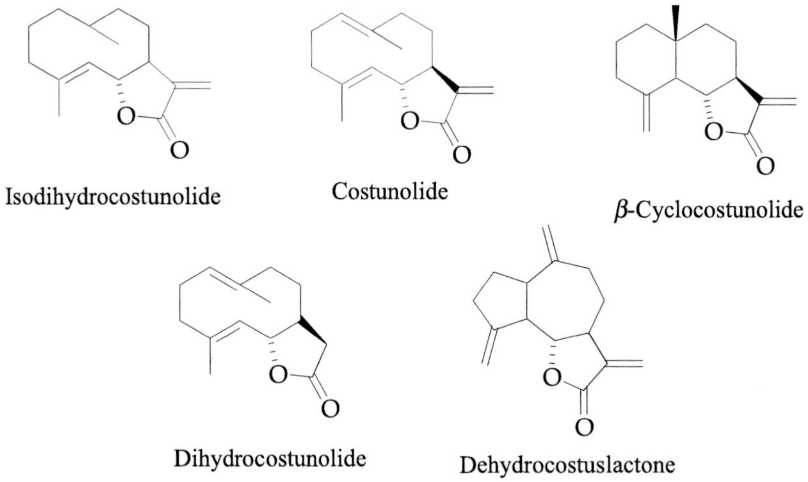

Isodihydrocostunolide

Costunolide

β-Cyclocostunolide

Dihydrocostunolide

Dehydrocostuslactone

Figure 4.6 Sesquiterpenes of *Saussurea lappa*.

4.7 MILLERENOLIDE AND THIELEANIN

Millerenolide and thieleanin (Fig. 4.7) isolated from *Viguiera sylvatica* and *Decachaeta thieleana,* respectively showed a similar pattern of cytotoxicity with the greatest effect on viability being evident with A549 human lung cancer cells IC50—40 and 32 μM respectively), and with the 3T3/HER2 cell line which are 3T3 mouse fibroblasts transfected with the HER2 oncogene IC50—16 and 28 μM respectively).

Treatment with millerenolide (8 mg/kg, IP on days 0, 2 and 4 post-inoculation) significantly inhibited the growth of subcutaneous B16/BL6 tumors in C57BL/6 mice, (50 percent inhibition at day 25, P = 0.015), as well as retarding the appearance of detectable tumor (millerenolide - day 15.2+/–0.4 vs control - day 12.8+/–0.5, mean+/–SEM, P = 0.011).[7]

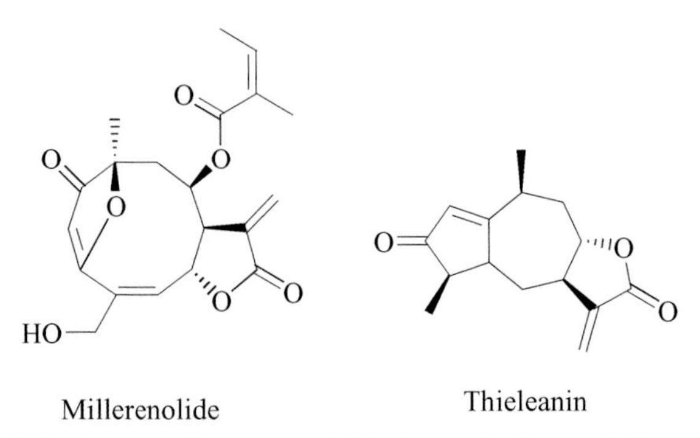

Figure 4.7 Cytotoxic sesquiterpenes of *Viguiera sylvatica* and *Decachaeta thieleana*.

4.8 DIHYDRO-β-AGAROFURAN SESQUITERPENES OF *CELASTRUS VULCANICOLA*

P-Glycoprotein overexpression is a contributing factor to multi-drug resistance in cancer cells and is one drawback in the treatment of cancer. Dihydro-β-agarofuran sesquiterpenes (Fig. 4.8) isolated from the leaves of *Celastrus vulcanicola* Donn. Sm. (Celastraceae) was assayed on human MDR1-transfected NIH-3T3 cells, in order to determine their ability to reverse the MDR phenotype due to P-Glycoprotein overexpression. Six of the isolated compounds showed an effectiveness that was similar to (or higher than) the classical P-Glycoprotein reversal agent verapamil for the reversal of resistance to daunomycin and vinblastine.[8]

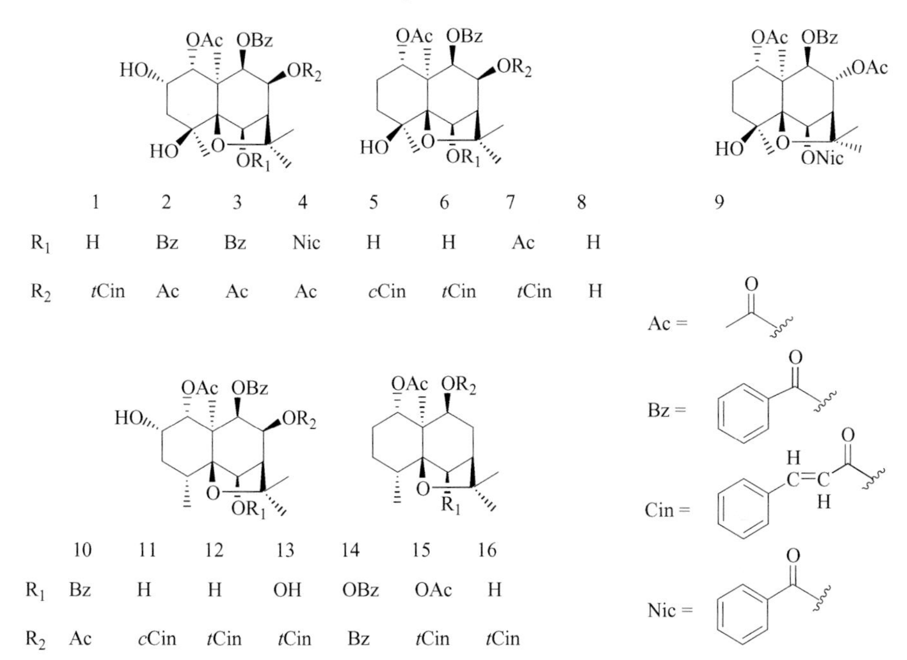

Figure 4.8 Dihydro-β-agarofuran sesquiterpenes of *Celastrus vulcanicola* Donn. Sm. (Celastraceae).

4.9 SANTAMARINE, 9β-ACETOXYCOSTUNOLIDE AND 9β-ACETOXYPARTHENOLIDE

Cyathocline purpurea (Buch.-Ham. ex D.Don) Kuntze (Asteraceae) has been traditionally used to treat various diseases including cancers for many years. Santamarine, 9β-acetoxycostunolide and 9β-acetoxyparthenolide inhibited the growth of L1210 murine leukaemia, CCRF-CEM human leukaemia, KB human nasopharyngeal carcinoma, LS174T human colon adenocarcinoma and MCF 7 human breast adenocarcinoma cells *in vitro*, with IC50 in the range of 0.16-1.3 µg/ml.

In L1210 model, santamarine and 9*β*-acetoxycostunolide (Fig. 4.9) inhibited L1210 cell growth, colony formation and [(3)H]-thymidine incorporation in time- and concentration-dependent ways. Flow cytometry studies showed that santamarine and 9*β*-acetoxycostunolide blocked L1210 cells in the G2-M phase of the cell cycle. DAPI staining and caspase activity assays showed santamarine and 9*β*-acetoxycostunolide induced apoptosis and activated caspase 3 in L1210 cells.[9]

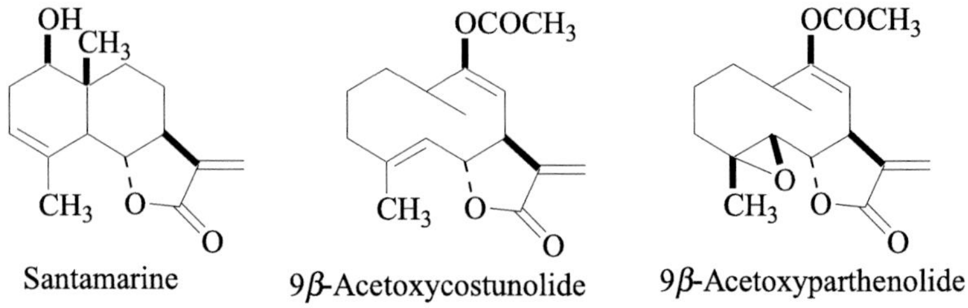

Santamarine 9*β*-Acetoxycostunolide 9*β*-Acetoxyparthenolide

Figure 4.9 Anticancer sesquiterpene lactones from *Cyathocline purpurea.*

4.10 FURANODIENE

Furanodiene (Fig. 4.10) is a sesquiterpene extracted from the essential oil of the rhizome of *Curcuma wenyujin* Y.H. Chen et C. Ling (Wen Ezhu). *In vitro*, MTT assay was used to compare the inhibitory effects of furanodiene and Wen Ezhu's essential oil on 11 human cancer cell lines. Compared to the essential oil, furanodiene showed stronger growth inhibitions on HeLa, Hep-2, HL-60, PC3, SGC-7901 and HT-1080 cells with IC50 between 0.6-4.8 μg/ml. *In vivo*, furanodiene exhibited inhibitory effects on the growth of uterine cervical (U14) and sarcoma 180 (S180) tumors in mice.[10]

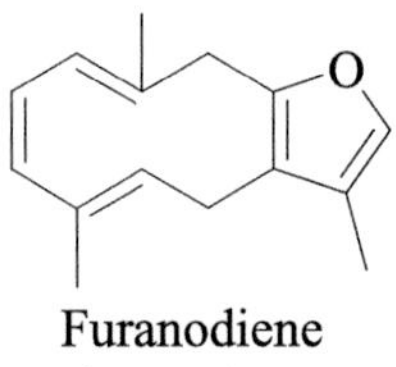

Furanodiene

Figure 4.10 Structure of sesquiterpene of *Curcuma wenyujin* Y.H. Chen et C. Ling.

4.11 GERMACRANE-TYPE SESQUITERPENOIDS FROM *MAGNOLIA KOBUS*

Cytotoxicity of germacrane-type sesquiterpenoids: costunolide, parthenolide, isobisparthenolidine and bisparthenolidine (Fig. 4.11), isolated from the chloroform-soluble fraction of the methanolic extract of the bark of *Magnolia kobus* DC (Magnoliaceae) were reported against human A549, SK-OV-3, SK-MEL-2, and HCT15 tumor cells.[11]

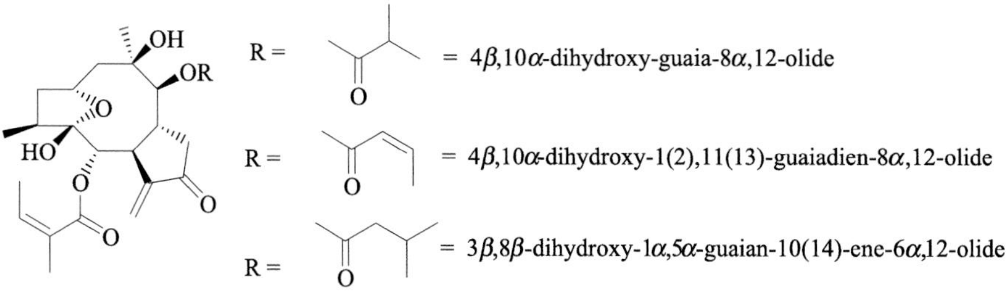

R = = 4β,10α-dihydroxy-guaia-8α,12-olide

R = = 4β,10α-dihydroxy-1(2),11(13)-guaiadien-8α,12-olide

R = = 3β,8β-dihydroxy-1α,5α-guaian-10(14)-ene-6α,12-olide

Figure 4.11 Germacrane-type sesquiterpenoids of *Magnolia kobus*.

4.12 CHLOROFORM EXTRACT OF *CARPESIUM ROSULATUM*

The chloroform extracts obtained from the whole plant of *Carpesium rosulatum* MIQ. (Asteraceae) exhibited significant anticancer activity against human tumor cell line, A549, SK-OV-3, SK-MEL-2, XF-498, and HCT-15.[12]

4.13 PINGUISANE-TYPE SESQUITERPENOIDS ISOLATED FROM *FRULLANIA* Sp. AND *PORELLA PERROTTETIANA*

Structure activity relationship studies showed that the presence of a phthalide group in bibenzyls, a α-methylene-β-lactone in germacrane-type sesquiterpenoids and β-hydroxycarbonyl in pinguisane-type sesquiterpenoids isolated from *Frullania* sp. and *Porella perrottetiana,* play an important role in cytotoxic activity against both human promyelocytic leukemia (HL-60) and human pharyngeal squamous carcinoma (KB) cell lines.[13]

4.14 COSTUNOLIDE

Investigatory research was undertaken to study mechanism of action of costunolide (Fig. 4.12) in chemotherapy for prostate cancer. Costunolide showed effective antiproliferative activity against hormone dependent (LNCaP) and independent (PC-3 and DU-145) prostate cancer cells. Costunolide induces apoptosis through nuclear calcium(2+) overload and DNA damage response in human prostate cancer.[14]

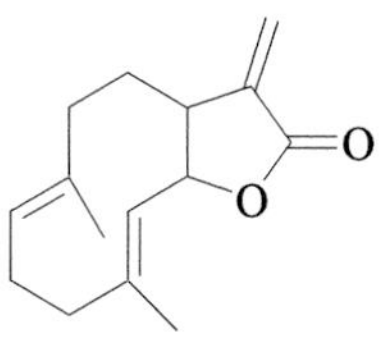

Costunolide

Figure 4.12 Structure of costunolide.

4.15 PARTHENOLIDE

The role of parthenolide (Fig. 4.13), a sesquiterpene lactone found in *Tanacetum parthenium* L. (Asteraceae) in three lines of human hepatocellular carcinoma cells was investigated. Co-treatment with parthenolide and TRAIL-induced apoptosis with the inactivation of caspases 8 and 3. This could be the basis for a novel therapeutic strategy for hepatic tumors.[15]

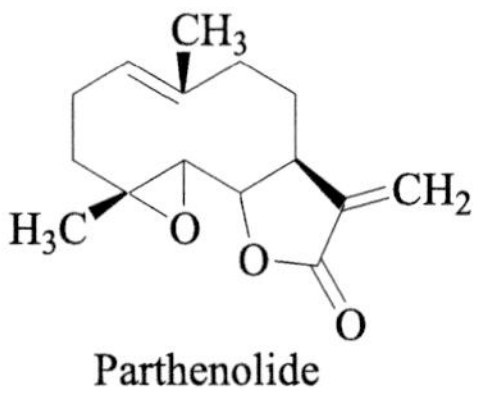

Parthenolide

Figure 4.13 Structure of sesquiterpene lactone found in *Tanacetum parthenium* L. (Asteraceae).

4.16 HIRSUTANOL A

The new hirsutane sesquiterpenoid, hirsutanol A (Fig. 4.14) isolated from AcOEt extract of the marine fungus *Chondrostereum* sp., (isolated from the soft coral *Sarcophyton tortuosum)* exhibited potent cytotoxic activities against various cancer cell lines.[16]

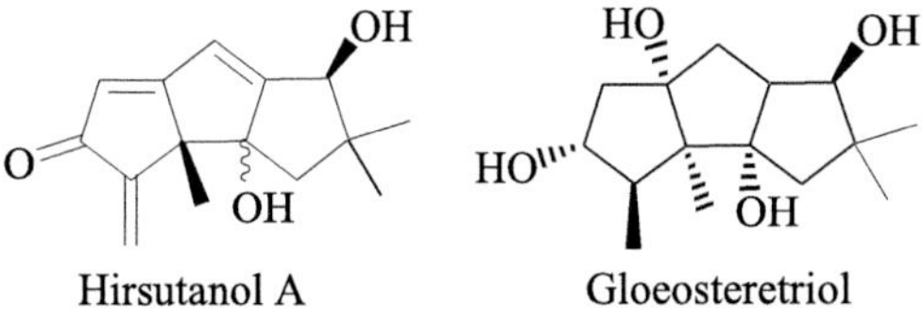

Hirsutanol A   Gloeosteretriol

Figure 4.14 Hirsutane sesquiterpenoids of *Chondrostereum* sp.

REFERENCES

1. Jie R, Jiang Y, Xiaonan Ma, *et al.* Antitumor Effects of Sesquiterpene Lactones from *Crepidiastrum sonchifolium* and their Derivates. *Asian J Trad Med* 2006; 1: 1-20.
2. Park HW, Choi SU, Baek NI, *et al.* Guaiane sesquiterpenoids from *Torilis japonica* and their cytotoxic effects on human cancer cell lines. *Arch Pharm Res* 2006; 29: 131-134.
3. Xu G, Liang Q, Gong Z, *et al.* Antitumor activities of the four sesquiterpene lactones from *Elephantopus scaber* L. *Exp Oncol* 2006; 28: 106-109.
4. Xie HG, Chen H, Cao B, *et al.* Cytotoxic germacranolide sesquiterpene from *Inula cappa. Chem Pharm Bull* (Tokyo) 2007; 55: 1258-1260.

5. Kim MR, Hwang BY, Jeong ES, *et al.* Cytotoxic germacranolide sesquiterpene lactones from *Carpesium triste* var. *manshuricum. Arch Pharm Res* 2007; 30: 556-560.
6. Robinson A, Kumar VT, Sreedhar E, *et al.* A new sesquiterpene lactone from the roots of *Saussurea lappa*: Structure–anticancer activity study. *Bioorg Med Chem Lett* 2008; 18: 4015-4017.
7. Taylor PG, Dupuy Loo OA, Bonilla JS, *et al.* Anticancer activities of two sesquiterpene lactones, millerenolide and thieleanin isolated from *Viguiera sylvatica* and *Decachaeta thieleana. Fitoterapia* 2008; 79: 428-432.
8. Torres-Romero D, Muñoz-Martínez F, Jiménez IA, Castanys S, Gamarro F, Bazzocchi IL. Novel dihy-dro-β-agarofuran sesquiterpenes as potent modulators of human P-glycoprotein dependent multidrug resistance. *Org Biomol Chem* 2009; 7: 5166-5172.
9. Ma G, Chong L, Li Z, *et al.* Anticancer activities of sesquiterpene lactones from *Cyathocline purpurea in vitro. Cancer Chemother Pharmacol* 2009; 64: 143-52.
10. Sun XY, Zheng YP, Lin DH, *et al.* Potential Anti-Cancer Activities of Furanodiene, A Sesquiterpene from *Curcuma wenyujin. Am J Chin Med* 2009; 37: 589-96.
11. Park HW, Lee JH, Choi SU, *et al.* Cytotoxic germacranolide sesquiterpenes from the bark of *Magnolia kobus. Arch Pharm Res* 2010; 33: 71-74.
12. Hyung-In Moon. Studies of the anticancer effect of sesquiterpene lactone from *Carpesium rosulatum. J Med Plant Res* 2010; 4: 1906-1909.
13. Komala I, Ito T, Nagashima F, *et al.* Cytotoxic bibenzyls, and germacrane- and pinguisane-type sesquiterpenoids from Indonesian, Tahitian and Japanese liverworts. *Nat Prod Comm* 2011; 6: 303-309.
14. Hsu JL, Pan SL, Ho YF, *et al.* Costunolide induces apoptosis through nuclear calcium 2+ overload and DNA damage response in human prostate cancer. *J Urol* 2011; 185: 1967-1974.
15. Carlisi D, D'Anneo A, Angileri L, *et al.* Parthenolide sensitizes hepato-cellular carcinoma cells to TRAIL by inducing the expression of death receptors through inhibition of STAT3 activation. *J Cell Physiol* 2011; 226: 1632-1641.
16. Li HJ, Lan WJ, Lam CK, *et al.* Hirsutane sesquiterpenoids from the marine-derived fungus *Chondrostereum* sp. *Chem Biodiv* 2011; 8: 317-324.

Chapter 5

Berberine—Alkaloid with Broad Spectrum of Pharmacological Activities

5.1 INTRODUCTION

Berberine (Fig. 5.1) is an isoquinoline alkaloid with a bright yellow color that is often seen in most herb materials which contain a significant amount of this compound. Berberine is the chief alkaloid from roots and stem-bark of *Berberis* species. It is found mostly from roots of *B. aristata, B. petiolaris, B. vulgaris, B. aquifolium, B. thunbergii, B. asiatica, Coptis teeta* and *Hydrastis canadensis* (see Table 5.1).[1-4]

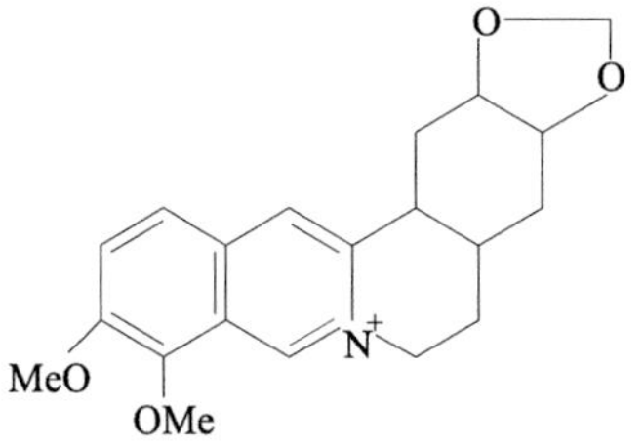

Figure 5.1 Chemical structure of berberine.

Table 5.1 Percentage of Berberine in various plant sources

Medicinal Plant	*Percent of Berberine*
B. aristata	5 percent in roots and 4.2 percent in stem-bark
B. aquifolium	-
B. petiolaris	0.43 percent in roots
B. thunbergii	-
Coptis teeta	8-9 percent in rhizomes
Hydrastis canadensis	-

5.2 BERBERINE FROM CHINESE MEDICINAL PLANTS

Among Chinese herbs, the primary sources are *B. sargentiana, Phellodendron amurense* and *Coptis chinensis. Coptis chinensis* rhizomes and related species used as its substitutes have about 4-8 percent berberine, while *Phellodendron amurense* bark has about half as much, at 2-4 percent berberine.[5]

5.3 PHARMACOLOGICAL STUDIES—PRE-CLINICAL

5.3.1 Anti-proliferative and Anti-migratory Activity

Berberine is capable of inhibiting growth and endogenous platelet-derived growth factor synthesis in vascular smooth muscle cells after *in vitro* mechanical injury. A study analyzed the effects of berberine on vascular smooth muscle cell growth, migration, and signaling events after exogenous platelet-derived growth factor stimulation *in vitro* in order to resemble a post-angioplasty platelet-derived growth factor shedding condition.

Pretreatment of vascular smooth muscle cells with berberine inhibited platelet-derived growth factor-induced proliferation. Berberine significantly suppressed platelet-derived growth factor F-stimulated cyclin D1/D3 and cyclin-dependent kinase gene expression. Moreover, berberine increased the activity of AMP-activated protein kinase, which led to phosphorylation activation of p53* and increased protein levels of the Cdk inhibitor p21**.

Compound C, an AMPK inhibitor, in part but significantly attenuated berberine-elicited growth inhibition. In addition, stimulation of vascular smooth muscle cells with platelet-derived growth factor led to a transient increase in GTP-bound, active form of Ras, Cdc42 and Rac1, as well as VSMC migration. However, pretreatment with berberine significantly inhibited platelet-derived growth factor F-induced Ras, Cdc42 and Rac1 activation and cell migration. Co-treatment with farnesyl pyrophosphate and geranylgeranyl pyrophosphate drastically reversed berberine-mediated antiproliferative and migratory effects in vascular smooth muscle cells. The observations offer a molecular explanation for the antiproliferative and antimigratory properties of berberine.[6]

5.3.2 Antimicrobial Activity

5.3.2.1 Antibacterial

In one experiment, berberine hydrochloride reduced the cholera toxin-induced secretion of water, sodium and chloride in perfused rat ileum. Berberine was also found to inhibit the intestinal secretory response of *Vibrio cholerae* and *Escherichia coli* enterotoxins without causing histological damage to the intestinal mucosa.[7]

Berberine is also active against other intestinal infections that cause acute diarrhea such as *Shigella dysenteriae, Salmonella paratyphi* and various

---

*Cellular tumar antigen   ** Cyclin-dependent kimass inhibitor1

Klebsiella species. Berberine sulfate has been shown to block the adherence of *Streptococcus pyrogenes* and *E. coli* to host cells, possibly explaining its mechanism of action against numerous pathogens.[8]

The effects of chlorpromazine (CPZ), berberine and verapamil on intestinal hyper secretion in the rabbit ileal loop model by the heat-labile enterotoxin (LT) of *Escherichia coli* were studied in relation to their ability to inhibit the stimulation of intestinal adenylate cyclase by the heat-labile enterotoxin. Chlorpromazine 5 mg by the intraluminal route and 4 mg/kg by the intramuscular route significantly reduced LT-induced intestinal hyper secretion. Berberine (10 mg) exerted an inhibitory effect, but only after IP administration, whereas verapamil did not exert any significant inhibitory effect when administered either (2.5 mg) or intramuscular (4 mg/kg). At concentrations of (0.17-1.34) × 10^{-3} M, Chlorpromazine, the anti-secretory effect of CPZ correlated with its inhibitory effect on rabbit heat-labile enterotoxin-stimulated intestinal adenylate cyclase. Inhibition of cAMP synthesis was probably not involved in the mechanism of action of the two other substances. These results indicate that chlorpromazine and phenothiazines in general are efficient drugs for reducing heat-labile enterotoxin-induced intestinal hyper-secretion and could represent a model for synthesis of new anti-secretory drugs with no tranquilizer side effects.[9]

Berberine was found to be the active constituent in an extract of *Hydrastis canadensis* root that demonstrated activity against a multiple drug-resistant strain of *Mycobacterium tuberculosis.*[10] Berberine is reported to inhibit *Helicobacter pylori.*[11]

The growth thermogenic curves of Escherichia coli affected by berberine, coptisine (Fig. 5.2) and palmatine (Fig. 5.3) extracted from *Coptis chinensis* were determined quantitatively by microcalorimetry. The power-time curves of *E. coli* with and without the three protoberberine alkaloids were acquired; at the same time the extent and dur ation of inhibitory effects on the metabolism were evaluated by growth rate constant (k), half-inhibitory ratio IC50, peak time of maximum heat-output power (tp), total heat-production (Qt). The inhibitory effects of three protoberberine alkaloids on *E. coli* revealed that the sequence of their antimicrobial activity was berberine > coptisine > palmatine.[12]

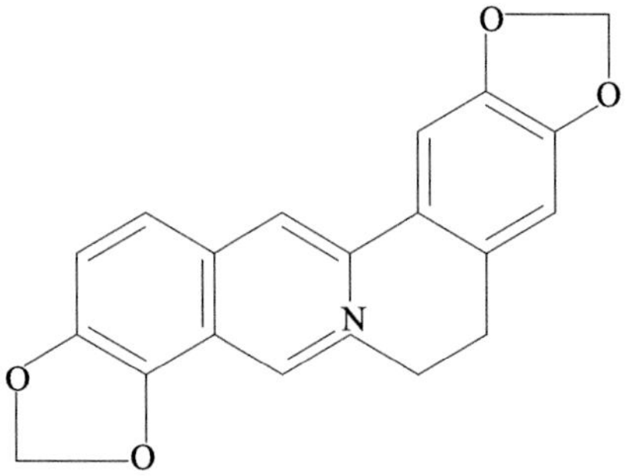

Figure 5.2 Chemical structure of coptisine.

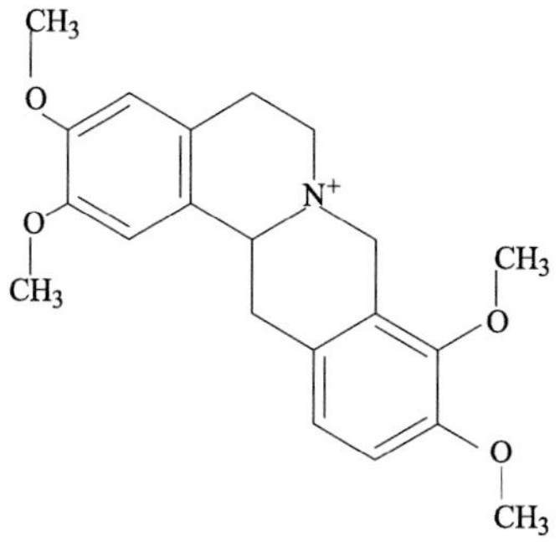

Figure 5.3 Chemical structure of palmitine.

Antibacterial activity of berberine is potentiated by methoxyhydnocarpin (Fig. 5.4). The observation led to the possibility that plants produce both antibacterial compounds and compounds which target bacterial efflux mechanisms to inhibit possible resistance to latent plant antibacterial in bacteria in their environment.[13]

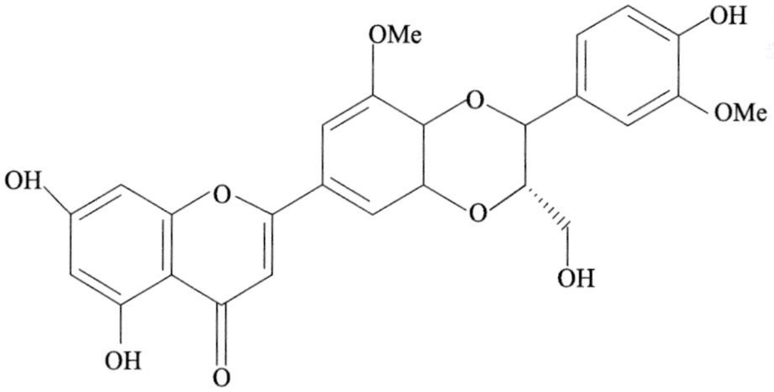

Figure 5.4 Structure of 5′ methoxyhydnocarpin.

5.3.2.2 Antifungal

The antifungal activity of trial denture cleansers prepared with berberine hydrochloride was examined against *Candida albicans, C. tropicalis* and *C. glabrata*. A commercial denture cleanser and a trial denture cleanser that exhibited strong antifungal activity were tested for their effects on *Candida* spp., the color stability of the dental material and the surface roughness of acrylic resin plates. The results of these tests revealed that the trial denture cleanser removed 64 to 89 percent of adhered cells from acrylic resin surfaces and had little effect on the other physical properties tested.[14]

5.3.2.3 Antiprotozoal

Parenteral administration of berberine has been shown to give rise to a statistically significant prolongation of the lives of rats infected with Trypanosoma equiperdum.[15] Berberine sulfate has been shown to inhibit the growth of *Entamoeba histolytica, Giardia lamblia* and *Trichomonas vaginalis in vitro.* The parasites all exhibited morphological changes after exposure to berberine sulfate.[16]

Earlier studies demonstrated the use of berberine in the treatment of *Leishmania donovani* infestation.[17-19] Berberine and several of its derivatives were tested for efficacy against *L. donovani* and *L. braziliensis* panamensis in golden hamsters. Tetrahydroberberine was less toxic and more potent than berberine against *L. donovani* but was not as potent as meglumine antimonate (Glucantime), a standard drug for the treatment of leishmaniasis. Only berberine and 8-cyanodihydroberberine showed significant activity (greater than 50 percent suppression of lesion size) against *L. braziliensis* panamensis.[20]

5.3.3 Gastrointestinal System

5.3.3.1 Cholagouge

Extract of *B. vulgaris* with 80 percent berberine and additional alkaloids stimulated the bile excretion of rats by 72 percent. Berberine has been shown to lower bilirubin levels.[21]

5.3.3.2 Anti-diarrheal activity

The motility of the small intestine in unanesthetized rats receiving berberine sulfate (0.2, 2.0, and 20.0 mg/kg IP) was investigated. Motility was determined by two methods—myoelectric activity was monitored with indwelling bipolar electrodes, and intestinal transit was measured by the movement of radio chromium (Na51CrO4). The 20.0 mg/kg dose caused a marked inhibition of spike activity for 21.8+/–7.0 min and disrupted activity fronts of the migrating myoelectric complex for 212.3 min.

Berberine, 2.0 mg/kg IP, disrupted migrating myoelectric complexes for 64.6 min but spike inhibition was not observed. Transit of the small intestine was significantly ($P < 0.001$) delayed at 15 and 100 min after the highest dose of berberine. Naloxone blocked the spike inhibition noted with 20.0 mg/kg of berberine but failed to improve the transit. Phentolamine blocked spike inhibition and was associated with a significantly earlier return of activity fronts of the migrating myoelectric complex.

Animals pretreated with this antagonist tended toward a higher geometric center in transit studies than those injected with berberine alone. Berberine was also administered by various routes (intraperitoneal injection, intravenous injection, orogastric gavage and intraluminal injection). An intraperitoneal injection was 10-fold more potent than an intravenous injection. Orogastric gavage and intraluminal administration of berberine did not alter intestinal motility. In summary, berberine sulfate significantly inhibits myoelectric activity and transit of the small intestine. This appears to be partially mediated by opioid and alpha-adrenergic receptors. The anti-diarrheal properties of berberine may be mediated, at least in part, by its ability to delay small intestinal transit.[22]

5.3.4 Hepatoprotective Activity

Hepatoprotective effect of Coptidis rhizoma aqueous extract and its possible mechanism were studied in rats intoxicated with carbon tetrachloride (CCl4) in the present study. SPRAGUE-Dawley (SD) rats at the age of 7 weeks were intraperitoneally injected with CCl4 at a dose of 1.0 ml/kg as a 50 percent olive oil solution. The rats were orally administered Coptidis rhizoma aqueous extract at doses of 400, 600, 800 mg/kg and 120 mg/kg berberine body weight after 6 hr of CCl4 treatment. At 24 hr after CCl4 injection, samples of blood and liver were collected and then biochemical parameters and histological studies were carried out.

The results showed that Coptidis rhizoma aqueous extract and berberine significantly inhibited the activities of alanine aminotransferase and aspartate aminotransferase and increased the activity of superoxide dismutase. Observation on the hepatoprotective effect of berberine was consistent to that of Coptidis rhizoma aqueous extract. The study demonstrated that Coptidis rhizoma aqueous extract has hepatoprotective effect on acute liver injuries induced by CCl4, and the results suggest that the effect of Coptidis rhizoma aqueous extract against CCl4-induced liver damage is related to anti-oxidant property.[23]

5.3.5 Cardiovascular System Activity

5.3.5.1 Anti-hypertensive activity

The alkaloid produces long lasting, dose related fall in blood pressure of anaesthetized rabbits.[1] Fractions from the root extracts of *B. vulgaris*, which contain 80 percent berberine and other alkaloids, have been shown to reduce the blood pressure of cats for several hours. With varying doses, both positive and negative inotropic effects on the cats' hearts were recorded.[24] Infusion of berberine when given intravenously to rats reduces blood pressure.[25]

5.3.5.2 Alpha 2 adrenoceptor antagonist activity

In this study, the interaction of berberine with human platelet alpha 2 adrenoceptor was investigated. Berberine was found to competitively inhibit the specific binding of [3H]-yohimbine. The displacement curve was parallel to those of clonidine, epinephrine, norepinephrine with the rank order of potency (IC50) being clonidine (0.4 μM) greater than epinephrine (7.5 μM) greater than norepinephrine (14.5 μM) = berberine (16.6 μM).

Increasing concentrations of berberine from 0.1 μM to 10 μM inhibited [3H]-yohimbine binding, shifting the saturation binding curve to the right without decreasing the maximum binding capacity. In platelet cyclic AMP accumulation experiments, berberine at concentrations of 0.1 μM to 0.1 μM inhibited the cAMP accumulation induced by 10 μM prostaglandin E1 in a dose dependent manner, acting as an alpha 2 adrenoceptor agonist. In the presence of L-epinephrine, berberine blocked the inhibitory effect of L-epinephrine behaving as an alpha 2 adrenoceptor antagonist. Three properties are similar to those of clonidine on human platelets, suggesting that berberine is a partial agonist of platelet alpha 2 adrenoceptors. These findings may account for hypotensive, anti-secretory, and sedative effects of berberine.[26]

5.3.5.3 Anti-arrhythmic activity

The study describes cardiovascular effects of berberine and its derivatives, tetrahydroberberine and 8-oxoberberine. Berberine has positive inotropic, negative chronotropic, anti-arrhythmic and vasodilator properties. Both derivatives of berberine have anti-arrhythmic activity. Some cardiovascular effects of berberine and its derivatives are attributed to the blockade of K+ channels (delayed rectifier and K (ATP)) and stimulation of Na+ –Ca (2+) exchanger. Berberine has been shown to prolong the duration of ventricular action potential. Its vasodilator activity has been attributed to multiple cellular mechanisms. The cardiovascular effects of berberine suggest its possible clinical usefulness in the treatment of arrhythmias and/or heart failure.[27]

5.3.5.4 Antiplatelet activity

In the present study, it was demonstrated *ex vivo* that berberine significantly inhibited rabbit platelet aggregation induced by adenosine diphosphate, arachidonic acid, collagen or calcium ionophore A23187. The most potent inhibition was observed in collagen-induced platelet aggregation. Using radioimmunoassay, we show *in vitro* that berberine significantly inhibited synthesis of thromboxane A_2 in rabbit platelets induced by adenosine diphosphate, arachidonic acid or collagen in which collagen-induced thromboxane A_2 synthesis was also most potently inhibited. In our *in vivo* study using radioimmunoassay, the plasma prostacyclin level was reduced by 34.6 percent during a 30-minutes period after intravenous administration of 50 mg/kg of berberine. The results suggest that berberine might inhibit arachidonic acid metabolism in rabbit platelets and endothelial cells at two or more sites—cyclooxygenase in the arachidonic acid cascade and possibly the enzyme(s) for arachidonic acid liberation from membrane phospholipid(s).[28]

5.3.5.5 Hypolipidemic activity

Berberine lowers elevated blood total cholesterol, LDL cholesterol triglycerides and aterogenic apolipoptoteins,[29] but the mechanism of action is distinct from statins.[30-33] Berberine reduces LDL cholesterol by upregulating LDLR mRNA expression posttranscriptionally while downregulating the transcription of proprotein convertase subtilisin/kexin type 9 (PCSK9), a natural inhibitor of LDL receptor and increasing in the liver the expression of LDL receptors through extra cellular signal-regulated kinase (ERK) signaling pathway[33] while statins inhibit cholesterol synthesis in the liver by blocking HMG-CoA-reduktase. This explains why berberine does not cause side effects typical of statins. Berberine activates AMP-activated protein kinase[34] specifically extrascellular signal-regulated kinases[35], which plays a central role in glucose and lipid metabolism[36] suppresses pro-inflammatory cystokines[37], and reduces MMP-9 and EMMPRIN expression[38], which are all beneficial changes for the health of the heart. Moreover, berberine reduces hepatic fat content in rats of non-alcoholic fatty liver disease). Berberine also prevents proliferation of hepatic stellate cells, which are central for the development of fibrosis during liver injury.[39]

5.3.6 Central Nervous System Activity

5.3.6.1 Anti-inflammatory activity

In the present study, the anti-inflammatory properties of total ethanol extract, three alkaloid fractions, a major alkaloid berberine and oxyacanthine (Fig. 5.5) isolated from *B. vulgaris* roots were compared. All these were applied in acute inflammation (carrageenan- and zymosan-induced paw oedema), as the total ethanol extract showed the highest reducing effect.

Their ability to alter *in vivo* and *in vitro* complementary activity was determined. Also, the total ethanol extract was most effective in a chronic inflammatory model of adjuvant arthritis. The protoberberine fractions (Bv2 and Bv3) and berberine suppressed a delayed type hypersensitivity (DTH) reaction. Fraction Bv1 and berberine diminished antibody response against SRBC *in vivo*. The *in vitro* treatment of splenocytes with berberine showed that the anti-SRBC antibody synthesis was influenced in a different manner depending on the time course of its application. Oxycanthine was less effective than berberine in the tests used.[40]

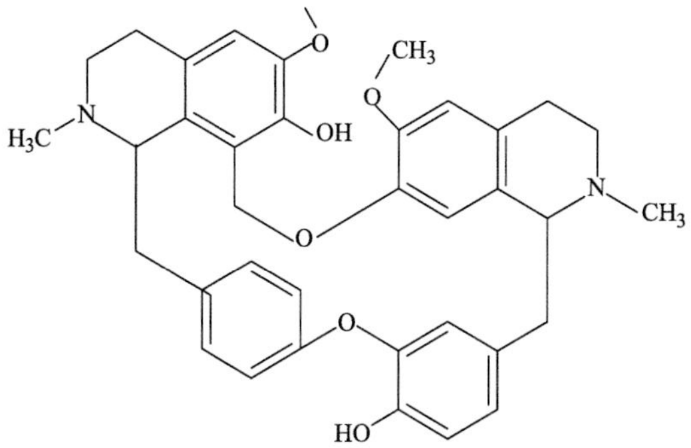

Figure 5.5 Structure of oxycanthine.

5.3.6.2 Antidepressant activity

A. The central depressant actions of methanol extract of coptis root, its active ingredients such as non-alkaloids fraction, tertiary base fraction, quarternary base fraction, magnoflorine fraction, berberine hydrochloride, coptisine hydrochloride and the extract from SAN O SHA SHIN TO (reparations which contain coptis root) were investigated in mice. The antigastric ulcer action of these substances was also examined in rats. All substances were given orally. Spontaneous movement and coordinative motor activity were not depressed by methanol extract, non-alkaloid fraction quarternary base fraction, magnoflorine fraction, berberine hydrochloride, coptisine hhdrochloride and the extract from SAN O SHA SHIN TO. There was no inhibition of chemical- and electro-shock-induced convulsion, morphine induced Straub's tail reaction, apomorphine-induced masticating motion and aggressive behavior induced by electrical stimulation. A loss of righting reflex* due to hypnotics was not potentiated by the substances. The quarternary base fraction did not elicit central depression, while the tertiary base fraction slightly depressed the function of the central nervous system. Quarternary base alkaloids such as berberine exerted a slight anti-ulcer effect.[41]

B. Berberine seems to act as an herbal antidepressant. Berberine inhibits prolyl oligopeptidase in a dose-dependent manner. Berberine is also known to bind sigma like many synthetic anti-depressant drugs. As berberine is a natural compound that has been safely administered to humans, preliminary results suggest the initiation of clinical trials in patients with depression, bipolar affective disorder, schizophrenia, or related diseases in which cognitive capabilities are affected, with either the extract or pure berberine.[42]

5.3.7 Genitourinary System

5.3.7.1 Reno protective

A study investigated the beneficial effects of berberine on renal function and its possible mechanisms in rats with diabetic nephropathy. Male Wistar rats were divided into three groups—normal, diabetic model and berberine treatment groups. Rats in the diabetic model and berberine treatment groups were induced to diabetes by intraperitonal injection with streptozotocin. Glomerular area, glomerular volume, fasting blood glucose, blood urea nitrogen, serum creatinine and urine protein for 24 hr were measured using commercially available kits. Meanwhile, the activity of superoxide dismutase content of malondialdehyde in serum, activity of aldose reductase and the expression of aldose reductase mRNA and protein in kidneys were detected by different methods.

The results showed that oral administration of berberine (200 mg/kg/day) significantly ameliorated the ratio of kidney weight to body weight Glomerular area, glomerular volume, fasting blood glucose, blood urea

*It is a reflex that corrects the orientation of the body when itis taken out of its normal upright position.

nitrogen, serum creatinine and urine protein for 24 hr were significantly decreased in the berberine treatment group compared with the diabetic model group ($P < 0.05$). Berberine treatment significantly increased serum SOD activity and decreased the content of MDA compared with diabetic model group ($P < 0.05$). Aldose reductase activity as well as the expression of aldose reductase mRNA and protein in the kidney was markedly decreased in the berberine treatment group compared with diabetic model group ($P < 0.05$).[43]

5.3.8 Anti-oxidant Activity

In a review, the effects of berberine on cultured rabbit corpus cavenosum smooth muscle cells damaged by hydrogen peroxide was studied through examining cell viability by methyl thiazolyl tetrazolium assay and assessing the level of malondialdehyde, superoxide dismutase activity, nitric oxide products and lactate dehydrogenase release in cells after stimulation with hydrogen peroxide.

Treatment with 1 mmol/L hydrogen peroxide significantly decreased the cell viability, nitric oxide products and superoxide dismutase activity of cultured rabbit corpus cavenosum smooth muscle cells from 100 percent to 48.57 ± 4.1 percent ($P < 0.01$), 66.8 ± 16.3 to 6.7 ± 2.1 μmol/L ($P < 0.01$), and 49.5 ± 1.8 to 30.1 ± 2.6 U/mL ($P < 0.01$), respectively, and increased lactate dehydrogenase release and malondialdehyde content from 497.6 ± 69.5 to 1100.5 ± 56.3 U/L ($P < 0.01$) and 3.7 ± 1.3 to 78.4 ± 2.9 nmol/mg protein ($P < 0.01$), respectively. However, treatment with different concentrations of Ber (10-1000 μmol/L) inhibited the damaging effects of hydrogen peroxide, with increased cell viability ($P < 0.05$ or $P < 0.01$), nitric oxide production ($P < 0.01$), superoxide dismutase activity ($P < 0.01$) and decreased lactate dehydrogenase release and malondialdehyde content (both $P < 0.01$).[44]

5.3.9 Absorption of Berberine

The aim of the present study was to use the P-glycoprotein inhibitors cyclosporin A, verapamil and the monoclonal antibody C219 in *in vivo* and *in vitro* models of intestinal absorption to determine the role of P-glycoprotein in berberine absorption. In the rat recirculating perfusion model, berberine absorption was improved 6-times by P-glycoprotein inhibitors. In the rat everted intestinal sac model, berberine serosal-to-mucosal transport was significantly decreased by cyclosporin A. In Ussing-type chambers, the rate of serosal-to-mucosal transport across rat ileum was 3-times greater than in the reverse direction and was significantly decreased by cyclosporin A. In Caco-2 cells, berberine uptake was significantly increased by P-glycoprotein inhibitors and by monoclonal antibody C219. P-glycoprotein appears to contribute to the poor intestinal absorption of berberine which suggests P-glycoprotein inhibitors could be of therapeutic value by improving its bioavailability.[45]

5.3.10 Clinical Studies

5.3.10.1 Oriental sore

Clinical studies have established the efficacy of hydrochloride of berberine in the treatment of oriental sore.[46]

5.3.10.2 Trachoma

Berberine has a long history of use for eye infections. In one study that looked at effectiveness in treating trachoma, berberine was more effective than sulfacetamide in eradicating *Chlamydia trachomatis* from the eye and preventing relapse of symptoms.[47]

5.3.10.3 Congestive heart failure

To determine the acute cardiovascular effects of berberine in humans, 12 patients with refractory congestive heart failure were studied before and during berberine intravenous infusion at rates of 0.02 and 0.2 mg/kg/min for 30 min. The lower infusion dose produced no significant circulatory changes, apart from a reduction in heart rate (14 percent).

The 0.2 mg/kg/min dose elicited several significant changes: (a) Decreases in systemic (48 percent, $P < 0.01$) and pulmonary vascular resistance (41 percent, $P < 0.01$), and in right atrium (28 percent, $P < 0.05$) and left ventricular end-diastolic pressures (32 percent, $P < 0.01$). (b) Increases in cardiac index (45 percent, $P < 0.01$), stroke index (45 percent, $P < 0.01$), and LV ejection fraction measured with contrast angiography (56 percent, $P < 0.01$). (c) Increases in hemodynamic and echocardiographic indices of LV performance: peak measured velocity of shortening (45 percent, $P < 0.01$), peak shortening velocity at zero load (41 percent, $P < 0.01$), rate of development of pressure at developed isovolumic pressure of 40 mmHg (20 percent, $P < 0.01$), percent fractional shortening (50 percent, $P < 0.01$), and the mean velocity of circumferential fiber shortening (54 percent, $P < 0.01$). (d) Decrease of arteriovenous oxygen difference (28 percent, $P < 0.05$) with no changes in total body oxygen uptake, arterial oxygen tension or hemoglobin dissociation properties.[48]

5.3.10.4 Hypercholesterolemia

It was recently reported that berberine lowers cholesterol through a mechanism different than that of statin drugs, suggesting a potential use both as an alternative to the statins and as a complementary therapy that might be used with statins in an attempt to gain better control over cholesterol. In a controlled Chinese study, it was shown that berberine, administered 500 mg twice a day for 3 mon, reduced serum cholesterol by 29 percent, triglycerides by 35 percent and LDL-cholesterol by 25 percent. The apparent mechanism increases the production of a receptor protein in the liver that binds the LDL-cholesterol, preparing it for elimination.[49]

5.3.10.5 Type 2 diabetes mellitus

A. In a study, evaluating efficacy of berberine in the treatment of diabetes mellitus, dietary therapy was first introduced to the patients for one month. For those who still had high fasting blood sugar, berberine was administered orally at a dose of 300, 400, or 500 mg each time, three times daily, adjusting the dosage according to the blood glucose levels; this treatment was followed for 1–3 months. A control group without diabetes was similarly treated, with no effect on blood sugar. For diabetic patients, it was reported that patients were less thirsty, consumed less water and urinated less, however had improved strength, and lower blood pressure; the symptoms declined in correspondence with declining blood glucose levels. Laboratory studies suggest that berberine may have at least two functions in relation to reducing blood sugar: inhibiting absorption of sugars from the intestine and enhancing production of insulin.[50]

B. Berberine has been shown to regulate glucose and lipid metabolism *in vitro* and *in vivo*. In a pilot study efficacy and safety of berberine in the treatment of type 2 diabetes mellitus patients was studied. In study A, 36 adults with newly diagnosed type 2 diabetes mellitus were randomly assigned to treatment with berberine or metformin (0.5 g 3 times a day) on a 3-months trial. The hypoglycemic effect of berberine was similar to that of metformin. Significant decreases in hemoglobin A (1c) were observed.[51]

5.3.11 Conclusion

Berberine has definite potential as a drug, since it possesses diverse pharmacological properties. Previous studies established the use of berberine as an antibacterial agent. However in recent studies, a striking effect of berberine is shown on the cardiovascular system. Since drug resistance and incidence of cardiovascular and metabolic diseases is on the rise, berberine holds promise for clinical trials.

REFERENCES

1. Watt G. *Dictionary of Economic Products of India,* Reprinted edition periodical expert. Delhi, Vol. VI (Pt. IV), 1972; pp. 83.
2. Nadkarni AK. *Nadkarni's Indian Materia Medica Popular Prakashan,* Bombay, 1976.
3. Chopra RN, Nayar SL, Chopra IC. *Glossary of Indian Medicinal Plants.* Council of Scientific and Industrial Research, New Delhi, 1996; pp. 247.
4. Gruenwald J, Brendler T, Jaenicke C. *PDR for Herbal Medicines.* Medical Economics Company, Montvale, NJ, 2000; pp. 313.

5. Dharmananda S. New uses of berberine, A Valuable Alkaloid from Herbs for "Damp-Heat" Syndromes. H:\NEW USES OF BERBERINE A Valuable Alkaloid from Herbs for "Damp-Heat" Syndromes.htm. 2005.
6. Liang KW, Yin SC, Tingh CT, *et al.* Berberine inhibits platelet-derived growth factor-induced growth and migration partly through an AMPK-dependent pathway in vascular smooth muscle cells. *Eur Pharmacol* 2008; 590: 343-354.
7. Sack RB, Froehlich JL. Berberine inhibits intestinal secretory response of Vibrio cholerae toxins and Escherichia coli enterotoxins. *Infect Immun* 1982; 35: 471-475.
8. Sun D, Courtney HS, Beachey EH. Berberine sulfate blocks adherence of Streptococcus pyogenes to epithelial cells, tibronectin, and hexadecane. *Antimicrob Agents Chemother* 1988; 32: 1370-1374.
9. Askri H, Scheftel JM. Effects of chlorpromazine, berberine and verapamil on *Escherichia coli* heat-labile enterotoxin-induced intestinal hyper secretion in rabbit ileal loops. *Med Microbiol* 1998; 27: 99-103.
10. Gentry EJ, Jampani HB, Keshavarz-Shokri A, *et al.* Antitubercular natural products: berberine from the roots of commercial *Hydrastis canadensis* powder. *J Nat Prod* 1988; 61: 1187-1193.
11. Bae EA, Han MJ, Kim NJ, *et al.* Anti-helicobacter pylori activity of herbal medicines. *Biol Pharm Bull* 1988; 21: 990-992.
12. Yan D, Jin C, Xiao XH, *et al.* Antimicrobial properties of berberine alkaloids in *Coptis chinensis* Franch by microcalorimetry. *J Biochem Biophys Meth* 2007; 70: 845-849.
13. Wagner H. Natural products chemistry and phytomedicine research in the new millennium: new developments and chalenges. *ARKIVOC* 2000; 27: 277-284.
14. Nakamoto K, Tamamoto M, Hamada T. *In vitro* study on the effects of trial denture cleansers with berberine hydrochloride. *J Prosthet Dent* 1995; 73: 530-533.
15. Serry TM, Bieter RN. A contribution of pharmacology of berberine. *Pharmacol Exp Ther* 1940; 69: 64-67.
16. Kaneda Y, Torii M, Tanaka T. *In vitro* effects of berberine sulfate on the growth of Entamoeba histolytica, Giardia lamblia and Tricomonas vaginalis. *Ann Trop Med Parasitol* 1991; 85: 417-425.
17. Hanson WL, Chapman WL, Jr, Kinnamon KE. Testing of drugs for antileishmanial activity in golden hamsters infected with Leishmania donovani. *Int J Parasitol* 1977; 7: 443-447.
18. Ghosh AK, Rakshit MM, Ghosh DK. Effect of berberine chloride on Leishmania donovani. *Indian J Med Res* 1983; 78: 407-416.
19. Ghosh AK, Bhattacharyya FK, Ghosh DK. Leishmania donovani: amastigote inhibition and mode of action of berberine. *Exp Parasitol* 1985; 60: 404-413.

20. Vennerstrom JL, Lovelace JK, Waits VB, *et al.* Berberine derivatives as antileishmanial drugs. *Antimicrob Agents Chemother* 1990; 34: 918-921.
21. Chan MY. The effect of berberine on bilirubin excretion in the rat. *Comp Med East West* 1977; 5: 161-168.
22. Eaker EY, Sninsky CA. Effect of berberine on myoelectric activity and transit of the small intestine in rats. *Gastroenterology* 1989; 96: 1506-1513.
23. Xingshen Ye, Feng Y, Tong Y, *et al.* Hepatoprotective effects of Coptidis rhizoma aqueous extract on carbon tetrachloride-induced acute liver hepatotoxicity in rats. *J Ethnopharmacol* 2009; 124: 130-136.
24. Lahiri SC. Positive and negative inotropic effects of berberine on the cats' heart. *Ann Biochem Exp Med* 1958; 18: 95.
25. Anonymous. *Wealth of India, Raw Materials*, Publications and Information Directorate, CSIR New Delhi, Vol. X, 1976; pp. 36.
26. Hui KK, Yu JL, Chan WF, *et al.* Interaction of berberine with human platelet alpha 2 adrenoceptors. *Life Sci* 1991; 49: 315-324.
27. Lau CW, Yao XQ, Chen ZY, *et al.* Cardiovascular actions of berberine. *Cardiovasc Drug Rev* 2001; 19: 234-244.
28. Huang G, Zhong CL, Shan W, *et al.* Effect of berberine on arachidonic acid metabolism in rabbit platelets and endothelial cells. *Thromb Res* 2002; 106: 223-227.
29. Zhou JY, Zhou SW, Zhang KB, *et al.* Chronic effects of berberine on blood, liver glucolipid metabolism and liver PPARs expression in diabetic hyperlipidemic rats. *Biol Pharm Bull* 2008; 31: 1169-1176.
30. Holy EW, Akhmedov A, Lüscher TF, *et al.* Berberine, a natural lipid-lowering drug, exerts prothrombotic effects on vascular cells. *J Mol Cell Cardiol* 2009; 46: 234-240.
31. Kong W, Wei J, Abidi P, *et al.* Berberine is a novel cholesterol-lowering drug working through a unique mechanism distinct from statins. *Nat Med* 2004; 10: 1344–1351.
32. Kim WS, Lee, YS, Cha SH, *et al.* Berberine improves lipid dysregulation in obesity by controlling central and peripheral AMPK activity. *Am J Physiol Endocrinol Metabol* 2009; 296: 812-819.
33. Abidi P, Zhou Y, Jiang JD, *et al.* Extra cellular signal-regulated kinase-dependent stabilization of hepatic low-density lipoprotein receptor mRNA by herbal medicine berberine. *Arterioscl Thromb Vas Biol* 2005; 25: 2170-2176.
34. Turner N, Li JY, Gosby A, *et al.* Berberine and its more biologically available derivative, dihydroberberine, inhibit mitochondrial respiratory complex I: a mechanism for the action of berberine to activate AMP-activated protein kinase and improve insulin function. *Diabetes* 2008; 57: 1414-148.
35. Lamontagne J, Pepin E, Peyot ML, *et al.* Pioglitazone acutely reduces insulin secretion and causes metabolic deceleration of the pancreatic {beta}-cell at submaximal glucose concentrations. *Endocrinology* 2009; 13: 213-128.

36. Lee YS, Kim WS, Kim KH, *et al.* Beberine, a natural plant product, activates AMP-activated protein kinase with beneficial metabolic effects in diabetic and insulin-resistant states. *Diabetes* 2006; 55: 2256-2264.
37. Jeong HW, Hsu KC, Lee JW, *et al.* Berberine suppresses proinflammatory responses through AMPK activation in macrophages. *Am J Physiol Endocrinol Metabol* 2009; 296: 955-964.
38. Huang Z, Wang L, Meng S, *et al.* Berberine reduces both MMP-9 and EMMPRIN expression through prevention of p38 pathway activation in PMA-induced macrophages. *Int J Cardiol* 2009; 23: 192-202.
39. Sun X, Zhang X, Hu H, *et al.* Berberine inhibits hepatic stellate cell proliferation and prevents experimental liver fibrosis. *Biol Pharm Bull* 2009; 32: 1533-1537.
40. Ivanovska N, Philipov S. Study on the anti-inflammatory action of *Berberis vulgaris* root extract, alkaloid fractions and pure alkaloids. *Int J Immunopharmacol* 1996; 18: 553-561.
41. Yamahara J. Behavioral pharmacology of berberine-type alkaloids. (1) Central depressive action of Coptidis rhizoma and its constituents. *Nippon Yakurigaku Zasshi* 1976; 72: 899-908.
42. Kulkarni SK, Dhir A. Current investigational drugs for major depression *Expert Opin Invest Drugs* 2009; 18: 767-788.
43. Liu WH, Hei ZQ, Nie H, Tang FT, Huang HQ, Li XJ, Deng YN, Chen SR, Guo FF, Huang WG, Chen FY, Liu PQ. Berberine ameliorates renal injury in streptozotocin-induced diabetic rats by suppression of both oxidative stress and aldose reductase. *Chinese Med J* 2008; 121: 706-712.
44. Tan Y, Tang Q, Hu B, *et al.* Antioxidant properties of berberine on cultured rabbit corpus cavernosum smooth muscle cells injured by hydrogen peroxide. *Acta. Pharmacol Sinica* 2007; 28: 1914-1918.
45. Pan Guo-yu, Wang G, Liu X, *et al.* The Involvement of P-Glycoprotein in Berberine Absorption. *Pharmcol Toxicol* 2008; 91: 193-197.
46. Dhar J. Berberine and oriental sore. *Ind J Dermatol Vernereol Lep* 1980; 46: 163.
47. Babbar OP, Chhatwal VK, Ray IB *et al.* Effect of berberine chloride eye drops on clinically positive trachoma patients. *Ind J Med Res* 1982; 76: 83-88.
48. Marin-Neto JA, Maciel BC, Seeches AL, *et al.* Cardiovascular effects of berberine in patients with severe congestive heart failure. *Clin Cardiol* 1988; 11: 253-260.
49. Weijia K, Wei J, Abidi P, *et al.* Berberine is a novel cholesterol-lowering drug working through a unique mechanism distinct from statins. *Nat Med* 2004; 10: 1344-1351.
50. Zhong Xi, Yi Jie, He Za Zhi. Therapeutic effect of berberine on 60 patients with non-insulin dependent diabetes mellitus and experimental research. Chinese *J Integ Trad West Med* 1995; 1: 91-95.
51. Yin J, Xing H, Ye J. Efficacy of berberine in patients with type 2 diabetes mellitus. *Alt Med Rev* 2008; 57: 712-717.

Chapter 6

Protopine—An Isoquinoline Alkaloid with Diverse Pharmacological Profile

6.1 INTRODUCTION

Protopine (Fig. 6.1) is an isoquinoline alkaloid that is usually found in most of the herb materials that contain any significant amount of this compound. It is the main alkaloid from roots and stems of *Corydalis* species. It is found mostly from *C. adunca, C. tashiro* and *C.govinaina.*[1-3] Other sources are *Papaver somniferum, Hypecoum lactiflorum, Chelidonium majus and Fumaria officinalis.*[4-7]

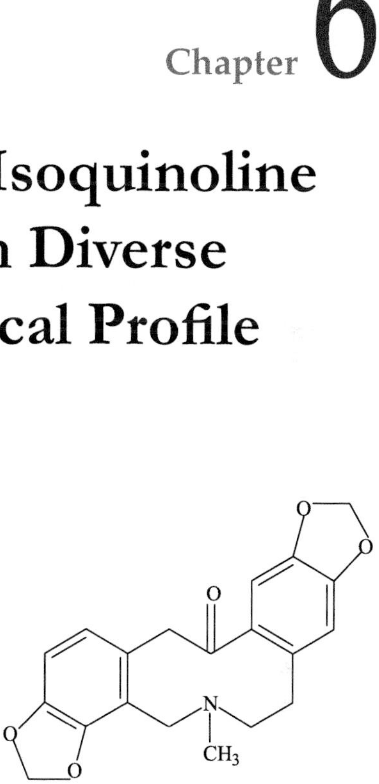

Figure 6.1 Chemical structure of protopine.

6.2 PHARMACOLOGY

6.2.1 Antithrombotic and Anti-inflammatory Activity

Five antiplatelet agents (apigenin, magnolol, osthole, protopine and norathyriol) isolated from Chinese herbs were evaluated for antihaemostatic and antithrombotic activity. Apigenin and magnolol are inhibitors of thromboxane synthesis, while osthole, protopine and norathyriol are inhibitors of phosphoinositide breakdown. Thirty minutes after intraperitoneal (IP) administration of these drugs, tail bleeding time of mice was prolonged markedly in a dose-dependent manner by norathyriol, protopine, osthole and magnolol, but not by apigenin. However, the antiplatelet agents (up to 200 mg/kg, IP) could not prevent acute thromboembolic death in mice. In endotoxin-induced experimental disseminated intravascular coagulation in rats, norathyriol (50-100 mg/kg, IP) prevented the decrease in platelet counts and fibrinogen, and the prolongation of plasma prothrombin time. Norathyriol (100 mg/kg, IP) also suppressed *ex-vivo* platelet aggregation induced by collagen and ADP in rat plasma.[8]

In a study, the effects of protopine on human platelet aggregation and arachidonic acid metabolism via cyclooxygenase and lipoxygenase enzymes were examined. Platelet aggregation induced by various platelet agonists was strongly inhibited by protopine in a concentration-related manner. The IC50 values (μM) of protopine (mean +/− SEM) against: arachidonic acid; 12 +/− 2: ADP; 9 +/− 2: collagen; 16 +/− 2 and platelet aggregation factor; 11 +/− 1, were much less than those observed for aspirin. In addition, protopine selectively inhibited the synthesis of thromboxane A2 via cyclooxygenase pathway. *In vivo*, pretreatment with protopine (50-100 μM/kg) protected rabbits from the lethal effects of arachidonic acid (2 μM/kg) or platelet aggregation factor (11 μM/kg) in dose-dependent manner. Protopine (50-100 μM/kg) also inhibited carrageenan-induced rat paw oedema with a potency of three-fold as compared to aspirin.[9]

Forty-one isoquinoline alkaloids were tested for antiplatelet aggregation effects. Among them, (–)-discretamine, protopine), ochotensimine, O-methylarmepavinemethine, lindoldhamine, isotetrandrine, thalicarpine, papaverine and D-(+)- N-norarmepavine exhibited significant inhibitory activity towards adenosine 5′-diphosphate (ADP)-, arachidonic acid, collagen and/ or platelet-activating factor platelet aggregation factor-induced platelet aggregation.[10]

6.2.2 Analgesic Activity

The analgesic effect of protopine was confirmed by tail-pinch and hot-plate tests when given subcutaneous (SC) 10-40 mg/kg, and 20-40 mg/kg inhibited the spontaneous movements of mice. Pro 40 mg/kg increased the sleeping rate, prolonged sleeping duration, and shortened the sleeping latency in mice hypnotized by IP pentobarbital sodium 30 mg/kg. Pro 10-40 mg/kg did not affect the inflammatory reaction induced by xylene and egg white. An icv injection of Pro 20-200 μM/mouse showed a remarkable analgesic effect in mice. The icv pretreatment of naloxone 2 μM blocked the analgesic effect completely. CaCl2 40 μM/mouse (ICV) or methotrexate 10 mg/kg (IP), an agonist of Ca2+ channel, showed a complete blockade of the analgesia, while nifedipine 100 mg/kg (PO), a blocker of Ca2+ channel, enhanced the analgesic effect. The IP pretreatment of reserpine 4 mg/kg reduced the Pro analgesia. Phentolamine 10 mg/kg (IP), a α-adrenergic blocker, tended to weaken the analgesia, but propranolol 10 mg/kg (IP), a β-blocker, did not affect it.[11]

6.2.3 Antispasmodic Activity

Two ethanolic dry extracts derived from *Chelidonium majus* Linn. with a defined content of the main alkaloids (chelidonine, protopine and coptisisine) and the alkaloids themselves were studied in three different antispasmodic test models on isolated ileum of guinea-pigs. In the BaCl2-stimulated ileum, chelidonine and protopine exhibited the known papaverine-like musculotropic action, whereas coptisine (up to $3.0 \times 10(-5)$ g/ml) was ineffective in this model. Both extracts were active with 53.5 and 49.0 percent relaxation at $5 \times 10(-4)$ g/ml. The carbachol and the electric field stimulated contractions were antagonized by all three alkaloids. Coptisine showed competitive antagonist behavior with a pA2 value of 5.95. Chelidonine and protopine exhibited a certain degree of non-competitive antagonism. In the electric field the antagonist activities decreased in the order protopine > coptisine > chelidonine. The concentrations of the chelidonium herb extracts for 50 percent inhibition of the carbachol and electrical field induced spasms were in the range of 2.5 to $5 \times 10(-4)$ g/ml.[12]

6.2.4 Anticholinesterase and Anti-amnesic Activities

In a study methanolic extract of the tuber of *Corydalis ternata* showed significant inhibitory effects on acetylcholinesterase. Further fractionation of this extract using acetylcholinesterase inhibition as the parameter screened resulted in the isolation and purification of an alkaloid, protopine. Protopine inhibited acetylcholinesterase activity in a dose-dependent manner. The concentration required for 50 percent inhibition was 50 μM. The anti-acetylcholinesterase activity of protopine was specific, reversible and competitive in manner. Furthermore, when mice were pretreated with protopine, the alkaloid significantly alleviated scopolamine-induced memory impairment. In fact, protopine had an efficacy almost identical to that of velnacrine, which is used to treat Alzheimer's disease, at an identical therapeutic concentration.[13]

Some authors have used protopine in the treatment of tremors. Protopine has an anticholinergic and GABA-ergic effect, and may have an effect similar to that of neuroleptics. Nevertheless, it is not clear whether the recommended quantity protopine is sufficient for a clinical effect.[14]

6.2.5 Anti-asthmatic Activity

A study was conducted for evaluating the anti-asthmatic potential of protopine. *In vitro* effects of protopine on the concentration response curves of guinea pig tracheal spiral strips for histamine and acetylcholine were observed. The contents of cAMP and cGMP in the rabbit's tracheal smooth muscle affected by protopine were determined by radioimmunoassay. The phosphodiesterase activities in guinea pig trachealis affected by protopine were assayed with reversed phase HPLC.

Protopine markedly depressed the constriction of guinea pig tracheal spiral strips induced by histamine or acetylcholine, and shifted the concentration response curves to the right. The maximal responses were significantly depressed, showing a non competitive antagonism. The pD 2′ of protopine for constrictions induced by histamine and acetylcholine were 3.11 and 3.45, respectively. Protopine increased cAMP levels in rabbit tracheal smooth muscles, but had no significant increase in their cGMP level. Protopine inhibited the cAMP phosphodiesterase activities in guinea pig tracheal smooth muscles, but had no significant decrease in their cGMP phosphodiesterase activities.[15]

6.2.6 Cardiovascular Activity

The aim of this study was to elucidate the ionic mechanisms of protopine effects in the heart. In single isolated ventricular myocytes from guinea-pigs, extracellular application of protopine markedly and reversibly abbreviates action potential duration, and decreases the rate of upstroke in a dose-dependent manner. Additionally, it produces a slight, but significant hyperpolarization of the resting membrane potential. Protopine at 25, 50 and 100 μM reduces L-type Ca2+ current (ICa,L) amplitude to 89.1, 61.9 and 45.8 percent of control, respectively, and significantly slows the decay kinetics of ICa, L at higher concentrations.

In the presence of protopine, both the inward rectifier (IK1) and delayed rectifier (IK) potassium currents were variably inhibited, depending on protopine concentrations. Sodium current (INa), recorded in low [Na+]o (40 μM) solution, is more potently suppressed by protopine At 25 μM, protopine significantly attenuated INa at most of the test voltages (–60 + 40 mV, with a 53 percent reduction at –30 mV). The results prove that protopine is not a selective Ca2+ channel antagonist but acts as a promiscuous inhibitor of cation channel currents including ICa, L, IK, IK1 as well as INa.[16]

A study was carried out to investigate the effects of protopine on K(ATP) channels and big conductance (BKCa) channels. Protopine concentration-dependently inhibited K(ATP) channel currents in human embryonic kidney cells (HEK-293) which were cotransfected with Kir6.1 and sulfonylurea receptor 1 (SUR1) subunits, but not with Kir6.1 cDNA transfection alone. At 25 μM, protopine reversibly decreased Kir6.1/SUR1 currents densities from –17.4 +/–3 to –13.2 +/–2.4 pA/pF at –60 mV ($n = 5$, $P < 0.05$). The hetero-logously expressed mSlo-encoded BK (Ca) channel currents in HEK-293 cells were not affected by protopine (25 μM), although iberiotoxin (100 μM) significantly inhibited the expressed BK (Ca) currents ($n = 5$, $P < 0.05$). To conclude, protopine selectively inhibited K(ATP) channels by targeting on SUR1 subunit.[17]

6.2.7 Neuroprotective Activity

In this study, the effect of protopine on the focal cerebral ischemia was investigated in rats. Male Sprague-Dawley rats were divided into five groups: sham-operated group, vehicle-treated group and three doses of protopine-treated groups (0.98, 1.96 and 3.92 mg/kg). Protopine was intraperitoneally administered to rats once daily for 3 days prior to the ischemia and 0.9 percent normal saline to rats in the vehicle-treated group in the same way. Rats in the sham-operated group were given 0.9 percent normal saline without the ischemia. The focal cerebral ischemia was induced by the middle cerebral artery occlusion for 24 hr via the intraluminal filament technique.

The results showed that pretreatment with protopine reduced the cerebral infarction ratio and serum lactate dehydrogenase activity, and improved the ischemia-induced neurological deficit score and histological changes of the brain in a dose-dependent manner. The studies further demonstrated that protopine increased superoxide dismutase activity in the serum, and decreased total calcium and terminal deoxynucleotidyl transferase-mediated dUTP nick end labeling (TUNEL)-positive cells in the ischemic brain tissue in the middle cerebral artery occlusion rats.[18]

In the present study, neuroprotective activity of protopine was studied against H(2)O(2)*-induced injury in PC12 cells. Pretreatment of PC12 cells with protopine improved the cell viability, enhanced activities of superoxide dismutase, glutathione peroxidase and catalase and decreased malondialdehyde level in the H(2)O(2) injured cells. Protopine also reversed the increased intracellular Ca(2+) concentration and the reduced mitochondrial membrane potential caused by H(2)O(2) in the cells. Furthermore, protopine was able to inhibit caspase-3 expression and cell apoptosis induced by H(2)O(2).[19]

6.2.8 Antidepressant Activity

Protopine isolated from *Dactylicapnos scandens* Hutch was identified as an inhibitor of both serotonin transporter and noradrenaline transporter *in vitro* assays. 5-hydroxy-DL-tryptophan (5-HTP)-induced head twitch response (HTR) and tail suspension test were adopted to study whether protopine has an antidepressant effect in mice using reference antidepressant fluoxetine and desipramine as positive controls. In the HTR test, protopine at doses of 5, 10, 20 mg/kg dose dependently increase the number of 5-HTP-induced HTR. Protopine at doses of 3.75 mg/kg, 7.5 mg/kg and 30 mg/kg also produces a dose-dependent reduction in immobility in the tail suspension test.[20]

6.2.9 Hepatoprotective Activity

This study demonstrates the hepatoprotective potential of 50 percent ethanolic water extract of the whole plant of *Fumaria indica* and its three

---

*Hydrogen peraoxide

fractions viz., hexane, chloroform and butanol against d-galactosamine induced hepatotoxicity in rats. The hepatoprotection was assessed in terms reduction in histological damage, changes in serum enzymes and metabolites bilirubin, reduced glutathione and lipid peroxidation. Among fractions more than 90 percent protection was found with butanol fraction in which alkaloid protopine was quantified as the highest i.e. about 0.2 μg/g by HPTLC. The isolated protopine in doses of 10-20 μg PO also proved equally effective hepatoprotective as the standard drug silymarin (single dose 25 μg PO). In general all treatments excluding hexane fraction proved hepatoprotective at par with silymarin.[21]

6.2.10 Antimicrobial Activity

6.2.10.1 Antiviral and antibacterial

In a study, 33 isoquinoline alkaloids belonging to protopine-, benzylisoquinoline-, benzophenanthridine-, spirobenzylisoquinoline-, phthalideisoquinoline-, aporphine-, protoberberine-, cularine- and isoquinolone-types as well as seven derivatives of them obtained from some Fumaria and Corydalis species growing in Turkey were evaluated for their *in vitro* antiviral and antimicrobial activities.

Both DNA virus Herpes simplex and RNA virus Para influenza were used for antiviral assessment of the compounds using Madine-Darby bovine kidney and Vero cell lines and their maximum non-toxic concentrations and cytopathogenic effects were determined using acyclovir and oseltamivir as references.

Antibacterial and antifungal activities of the alkaloids were tested against *Escherichia coli, Pseudomonas aeruginosa, Proteus mirabilis, Klebsiella pneumoniae, Acinetobacter baumannii, Staphylococcus aureus, Bacillus subtilis* and *Candida albicans* by the microdilution method and compared to ampicilline, ofloxacine and ketocanazole as references. The alkaloids did not present any notable antibacterial effect, while they had significant antifungal activity at 8 μg/ml concentration. On the other hand, the alkaloids were found to have selective inhibition against the para-influenza virus ranging between 0.5 and 64 μg/ml as minimum and maximum cytopathogenic effects inhibitory concentrations, whereas they were completely inactive towards Herpes simplex.[22]

6.2.10.2 Antiprotozoal

Four alkaloids, protopine, scoulerine, cheilanthifoline and stylopine, isolated from the Bhutanese medicinal plant *Corydalis calliantha* Long, were tested for antimalarial activity. Protopine, and cheilanthifoline, showed promising *in vitro* antiplasmodial activities against Plasmodium falciparum, both wild type (TM4) and multidrug resistant (K1) strains with IC50 values in the range of 2.78-4.29 μM. The results support, at a molecular level, the clinical use of the medicinal plant in the treatment of malaria.[23]

REFERENCES

1. Tang YL, Yang AM, Zhang YS, *et al.* Studies on the alkaloids from the herb *Corydalis adunca. Zhongguo ZhongYao Za Zhi* 2005; 30: 195-197.
2. Chen JJ, Duh CY, Chen IS. New tetrahydroprotoberberine N-oxide alkaloids and cytotoxic constituents of *Corydalis tashiroi. Planta Med* 1997; 65: 643-647.
3. Watt G. *Dictionary of Economic products of India,* Reprinted edition periodical expert. Delhi, Vol. VI (Pt. IV), pp. 17.
4. Nadkarni AK. *Nadkarni's Indian Materia Medica,* Popular Prakashan, Bombay, 1976; pp. 33.
5. Istatkova PR, Denkova P, Dangaa S, *et al.* Alkaloids from Mongolian species *Hypecoum lactiflorum* Kar. et Kir., Pazij. *Nat Prod Res* 2009; 23: 982-987.
6. Taborska E. The alkaloids of *Chelidonium majus* L. and their variability. *Planta Med* 1996; 62, Abstracts of the 44th Ann Congress of GA, 145.
7. Manske RHF, Holmes HL. *The Alkaloids,* Vol. IV, L. Holmes, (Eds.). Academic Press, New York, 1954; pp. 157-159.
8. Teng CM, Ko FN, Wang JP, *et al.* Antihaemostatic and antithrombotic effect of some antiplatelet agents isolated from Chinese herbs. *J Pharm Pharmacol* 1991; 43: 667-669.
9. Saeed SA, Gilani AH, Majoo RU, *et al.* Anti-thrombotic and anti-inflammatory activities of protopine. *Pharmacol Res* 1997; 36: 1-7.
10. Chia YC, Chang FR, Wu CC, *et al.* Effect of isoquinoline alkaloids of different structural types on antiplatelet aggregation *in vitro. Planta Med* 2006; 72: 1238-1241.
11. Xu O, Jin RL, Wu YY. Opioid, calcium, and adrenergic receptor involvement in protopine analgesia. *Zhongguo Yao Li Xue Bao* 1993; 14: 495-500.
12. Hiller KO, Ghorbani M, Schilcher H. Antispasmodic and relaxant activity of chelidonine, protopine, coptisine and Chelidonium. *Planta Med* 1998; 64: 758-60.
13. Kim SR, Hwang SY, Jang YP, *et al.* Protopine from *Corydalis ternata* has anticholinesterase and antiamnesic activities. *Planta Med* 1999; 65: 218-221.
14. Rektorová RI, Suchv V. How to treat tremor. *J Neurol* 2004; 251: 525-528.
15. Hua T. Effects of protopine on tracheal smooth muscle and its mechanism summary. *Chin J Pharmacol Toxicol* 2007; 32: 77-79.
16. Song LS, Ren GJ, Chen ZL, *et al.* Electro-physiological effects of protopine in cardiac myocytes inhibition of multiple cation channel currents. *Br J Pharmacol* 2000; 129: 893-900.
17. Jiang B, Cao R, Wang R. Inhibitory effect of protopine on K [ATP] channel subunits expressed in HEK-293c. *Eur J Pharmacol* 2004; 506: 93-100.

18. Xiao X, Liu J, Li T, *et al.* Protective effect of protopine on the focal cerebral ischaemic injury in rats. *Basic Clin Pharmacol Toxicol* 2007; 101: 85-89.
19. Xiao X, Liu J, Hu J, *et al.* Protective effects of protopine on hydrogen peroxide-induced oxidative injury of PC12 cells via Ca (2+) antagonism and antioxidant mechanisms. *Eur J Pharmacol* 2008; 591: 21-27.
20. Xu LF, Chu WJ, Qing XY, *et al.* Protopine inhibits serotonin transporter and noradrenaline transporter and has the antidepressant-like effect in mice models. *Neuropharmacology* 2006; 50: 934-940.
21. Rathi A, Srivastava AK, Shirwaikar A, *et al.* Hepatoprotective potential of *Fumaria indica* Pugsley whole plant extract, fractions and an isolated alkaloid protopine. *Phytomedicine* 2008; 15: 470-477.
22. Orhana I, Ozcelik B, Karaoğlu T, *et al.* Antiviral and antimicrobial profiles of selected isoquinoline alkaloids from Fumaria and Corydalis species. *Z Naturforsch* 2007; 62: 19-26.
23. Wangchuk P, Bremner JB, Rattanajak R, *et al.* Antiplasmodial agents from the Bhutanese medicinal plant *Corydalis calliantha. Phytother Res* 2010; 24: 481-485.

Chapter 7

Piperine—Review of Advances in Pharmacology

7.1 INTRODUCTION

In recent years many researchers have examined the effects of plants used traditionally by indigenous healers and herbalists to support the function and treatment of diseases. In most cases, scientists have confirmed the veracity of traditional experience and wisdom by discovering the mechanism of action of these plants. *Piper longum* L. and *Piper nigrum* L. (Piperaceae) are used in Indian traditional medicine and as a spice globally.[1]

Piperine (Fig. 7.1), an alkaloid is responsible for the pungency of *P. nigrum* L. and *P. longum* L.[1] Piperine can be obtained from the oleoresin in peppercorns. Piperine consists of about 5-7 percent of peppercorns. It exhibits a wide variety of biological effects.

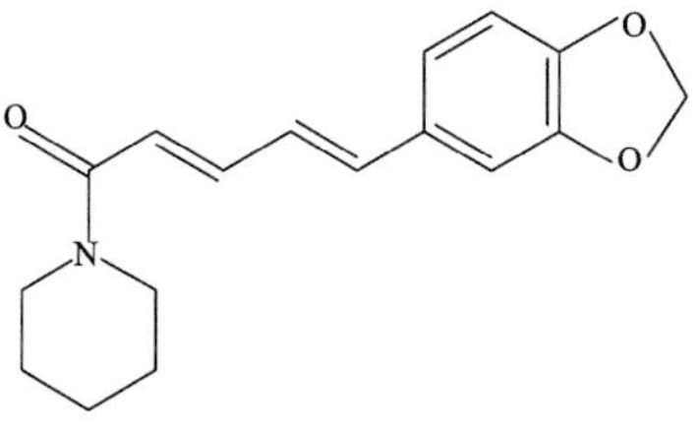

Figure 7.1 Chemical structure of piperine.

7.2 PHARMACOLOGY

7.2.1 Antidepressant Activity

A. In a study Song et al., investigated the antidepressant effect of piperine in mice exposed to chronic mild stress procedure. Repeated administration of piperine for 14 days in doses of 2.5, 5 and 10 mg/kg reversed chronic stress, induced changes in sucrose consumption, plasma corticosterone level and open field activity. Furthermore, the decreased proliferation of hippocampal progenitor cells was ameliorated and the level of brain-derived neurotrophic factor in hippocampus of chronic stressed mice was upregulated by piperine treatment.[2]

B. In another study, Wattanathorn et al., administered piperine to Wister male rats, at various doses ranging from 5, 10 and 20 mg/kg/day, body weight, (PO) for 4 weeks and the neuropharmacological activity (elevated plus maze, spontaneous locomotors behavior, forced swimming test, cognitive function) was determined after single, 1, 2, 3 and 4 weeks of treatment. The results showed that piperine during the entire dosage range possessed anti-depression activity and cognitive enhancing effect during the entire treatment duration.[3]

Kulkarni et al., evaluated the simultaneous administration of piperine (2.5 mg/kg, IP) with curcumin (20 and 40 mg/kg, IP) which resulted in the potentiation of antidepressant activities.[4]

7.2.2 Anti-oxidant

A. Zhao et al., noted the reversal of oxidative stress and hepatorenal dysfunction induced by beryllium. They observed that individual administration of gallic acid (50 mg/kg, IP) and piperine (10 mg/kg, PO) moderately reversed the altered biochemical variables, whereas the combination of these was found to completely reverse the beryllium-induced biochemical alterations and oxidative stress consequences. They concluded that gallic acid exerts a synergistic effect when administered with piperine.[5]

B. Nirala et al., evaluated the effect of piperine (10 mg/kg, 5 consecutive days, PO) individually and in combination with tiferron (300 mg/kg, IP) against beryllium (1 mg/kg/day, 28 days, IP) induced biochemical alteration and oxidative stress. They found that the combination of tiferron with piperine could reverse all the variables significantly towards the control.[6]

7.2.3 Bioenhancer Activity

A. Piperine and β-carotene

Vladimir et al., studied the effect of simultaneous administration of piperine (5 mg) on serum concentration of β-carotene (15 mg) in healthy volunteers for 14 days. The results indicate that there was a significant increase ($P < 0.0001$) in serum β-carotene concentration when supplemented with piperine (49.8 ± 9.6 µg/dL vs 30.9 ± 5.4 µg/dL) compared to β-carotene plus placebo, respectively. There was a 60 percent increase in the area under curve of β-carotene plus piperine when compared with β-carotene plus placebo.[7]

B. Piperine and thiobarbituric acid

Vijaykumar et al., studied the effect of simultaneous administration of piperine (0.02 g/kg, body weight) plus high-fat diet (containing 20 percent coconut oil, 2 percent cholesterol, and 0.125 percent bile salt) for 10 days on levels of thiobarbituric acid reactive substances, conjugated dienes and activities of superoxide dismutase, catalase, glutathione peroxidase, glutathione-S-transferase and reduced glutathione in the liver, heart, kidney, intestine and aorta were observed in rats. They concluded that simultaneous supplementation of a high fat diet with piperine lowered thiobarbituric acid reactive substances, conjugated dienes and maintained activities of superoxide dismutase, catalase, glutathione peroxidase, glutathione-S-transferase and reduced glutathione near those of control rats. Selvendiran et al., 2005b examined the protective effect of piperine on DNA damage and activities of detoxifying enzyme such as glutathione transferase, quinone reductase and UDP- glucuronosyl transferase in lung cancer bearing animals induced by Benzo (a) pyrene. They observed that supplementation of piperine (50 mg/kg, body weight) enhanced the activities of detoxification enzymes and reduced DNA damage as determined by single cell electrophoresis.[8]

C. Piperine and coenzyme Q10

Vladimir et al., (2000) studied the relative bioavailability of 90 and 120 mg of coenzyme Q10 simultaneous administered with piperine (5 mg) or a placebo in healthy adult male volunteers in a single-dose experiment or in separate experiments for 14 and 21 days. The result of the single and the 14th day dose study indicated smaller, but no significant increase in plasma concentration when compared with coenzyme Q10 plus the placebo. Supplementation of 120 mg coenzyme Q10 with piperine for 21 days produces a statistical difference ($P = 0.0348$), approximately 30 percent greater, area under the plasma curve than the coenzyme Q10 plus a placebo.[9]

D. Piperine and carbamazepine

Pattanaik et al., evaluated the effect of simultaneous administration of piperine (20 mg PO) on plasma concentration of carbamazepine (300 mg or 500 mg) twice daily in epileptic patients. They observed that piperine significantly increased the mean plasma concentrations of carbamazepine in both dose groups. There was a significant increase in AUC (0-12 hr) ($P < 0.001$), average C (ss) ($P < 0.001$), t (1\2el) ($P < 0.05$) and a decrease in K (el) ($P < 0.05$), in both the dose groups, whereas changes in K (a) and t(1\2a) were not significant. C max ($P < 0.01$) and t (max) ($P < 0.01$) were increased only in the 500 mg dose group.[10]

E. Piperine and nevirapine

Kasibhatta et al., administered piperine or a placebo to healthy adult males for 6 days. On day 7 piperine or a placebo was administered with nevirapine (200 mg). Blood samples were collected from 1 to 144 hr post-dose. They postulated that the mean maximum plasma concentration, area under the plasma concentration-time curve from 0 hr to the last measurable concentration, Area Under Curve extrapolated to infinity and C (last) values of nevirapine were increased by approximately 120, 167, 170 and 146 percent, respectively, when co-administered with piperine.[11]

F. Piperine and curcumin

Durgaprasad et al., evaluated the effect of oral curcumin (500 mg) with piperine (5 mg) on the pain, and the markers of oxidative stress in patients with tropical pancreatitis for 6 weeks. There was a significant reduction in the erythrocyte malonyldialdeyde levels following curcumin therapy compared with a placebo, with a significant increase in glutathione levels. Lambert et al., (2004) reported that piperine (70.2 μM/kg, PO) co-administered with (–)-Epigallocatechin-3-gallate (163.8 μM/kg, PO) to male CF-1 mice increased the plasma C(max) and area under the curve by 1.3-fold compared to mice treated with Epigallocatechin-3-gallate only. This appears to be due to inhibiting glucuronidation and gastrointestinal transit.[12]

7.2.4 Apoptosis Inhibition

Choi et al., demonstrated that piperine (10-100 μM) protect House Ear Institute-Organ of Corti-1 cells against cisplatin-induced apoptosis through the induction of heme oxygenase-1 expression in a dose- and time-dependent manner. The c-Jun N-terminal kinase pathway played an important role in piperine-induced heme oxygenase-1 expression.[13]

7.2.5 Genotoxicity

Selvendiran et al., revealed significant suppression (33.9-66.5 percent) in the micronuclei formation induced by benzo (a) pyrene and cyclophosphamide which was reduced following oral administration of piperine at doses of 25, 50 and 75 mg/kg in mice.[14]

7.2.6 Immunosuppression

Neelima et al., investigated the role of piperine in cadmium induced immuno-compromised murine splenocytes. The addition of piperine in various concentrations (1, 10 and 50 μg/ml) ameliorated oxidative stress markers, Bcl-2 protein expression, mitochondrial membrane potential, caspase-3 activity, DNA damage, splenic B and T cell population, blastogenesis and cytokines. The highest dose of piperine could completely abrogate the toxic manifestations of cadmium and the splenic cells behaved in similar way as the control cells.[15]

7.2.7 Anti-platelet Activity

Park et al., isolated four acidamides (piperine, pipernonaline, piperocta-decalidine, and piperlongumine) from fruits of *Piper longum* L. and examined the inhibitory effect on washed rabbit platelet aggression induced by collagen, arachidonic acid, platelet activating factor and thrombin. They showed dose dependent inhibitory activities on platelet aggression except for that induced by thrombin.[16]

7.2.8 Anti-inflammatory Activity

A. Sarvesh et al., demonstrated that piperine inhibits adhesion of neutrophils to endothelial monolayer due to its ability to block the tumor necrosis factor-α induced expression of cell adhesion molecules, i.e., intercellular adhesion molecule-1, vascular cell adhesion molecule-1 and E-selectin. They observed that pretreatment of endothelial cells with piperine blocks the phosphorylation and degradation of IκBα* by attenuating tumor necrosis factor-α induced IκB kinase activity.[17]

B. Pradeep et al., observed that piperine at 2.5, 5 and 10 μg/ml concentration inhibited the collagen matrix invasion of B16F-10 melanoma cells in a dose-dependent manner. It also significantly reduced the pro-inflammatory cytokines (such as IL-1β, IL-6, TNF-α, GM-CSF).[18]

7.2.9 Antihypertensive Activity

Taqvi et al., observed that intravenous administration of piperine caused a dose-dependent (1 to 10 mg/kg) decrease in mean arterial pressure in normotensive anesthetized rats; the next higher dose (30 mg/kg) did not cause any further change in mean arterial pressure. Piperine, *in vitro* study on rabbit heart causes a partial inhibition of force, rate of contraction and coronary flow. In rabbit aortic ring, piperine inhibited high K+ (80 μM) precontractions and partially inhibited phenylephrine, due to Ca2+ channel blockade. In Ca2+-free medium, piperine (1 to 30 μM) exhibited vasoconstrictor effect.[19]

7.2.10 Hepatoprotective Activity

Matsuda et al., isolated piperine of ethyl acetate-soluble fraction from methanolic extract and concluded that piperine, dose-dependently inhibited increase in serum GPT and GOT levels at doses of 2.5-10 mg/kg (PO) in D-galactosamine induced liver toxicity in mice, and suggested that this inhibitory effect depended on the reduced sensitivity of hepatocytes to tumor necrosis factor-α.[20]

---

*It is part of the upstream NF-κB signal transduction cascade.

7.2.11 Anti-thyroid Activity

A. Vijayakumar et al., concluded that when piperine (40 mg/kg) was simul-taneously administered with carbimazole (10 mg) for 10 days significant reduction in plasma lipids and lipoproteins levels occurred, except for high density lipoprotein, which was significantly elevated. Piperine supplementation also improved the plasma levels of apo A-I, T3, T4, testosterone, and I and significantly reduced apo B, TSH, and insulin to near normal.[21]

B. Panda et al., administered piperine for 15 days at 0.25 and 2.50 mg/kg/day, (PO) to adult male Swiss albino mice. They concluded that piperine at dose (2.50 mg/kg) lowered the serum levels of both the thyroid hormones, thyroxin (T4) and triiodothyronine (T3) as well as glucose concentrations with a concomitant decrease in hepatic 5′D enzyme and glucose-6-phospatase (G-6-Pase) activity. However, no significant alterations were observed in animals treated with 0.25 mg/kg of piperine in any of the activities studied except an inhibition in serum T3 concentration.[22]

7.2.12 Fertility

Enhancer Pawinee et al., evaluated the effect of piperine on the fertilization of eggs in female hamsters from day 1 through day 4 of the oestrous cycle at dose of 50 and 100 mg/kg (body weight, PO). They observed that there was enhancement of fertilization, 85.4 ± 4.1 and 82.8 ± 4.8 at doses of 50 and 100 mg/kg, respectively at 9 hr after artificial inseminated. However, examination of the embryos retrieved 48 hr after artificial insemination revealed no difference in the stage of embryonic development.[23]

7.2.13 Anti-tumor Activity

A. Manoharan et al., investigated the chemoprotective effect of piperine (50 mg/kg, body weight, PO, alternate days) against 7,12-dimethylbenz[a] anthracene (0.5 percent in liquid paraffin, three times a week for 14 weeks) induced buccal pouch carcinoma of Syrian golden hamsters. They observed that piperine completely prevented the formation of oral carcinoma, probably due to its antilipidperoxidative and anti-oxidant potential as well as its modulating effect on the carcinogen detoxification process.[24]

B. Duessel et al., observed that piperine displayed an antiproliferation effect at 24 hr and statistically significant inhibition at 48 and 72 hr at 100-200 μM concentration against cultured human colon cancer cells (DLD-1).[25]

C. Wongpa et al., investigated the influence of piperine on chromosomes in rat bone marrow. Piperine administered to Wister male rats at dose of 100, 400 and 800 mg/kg, body weight, (PO) for 24 hr then challenged with cyclophosphamide at a dose of 50 mg/kg, body weight, (IP). They demonstrated that piperine at a dose of 100 mg/kg, gave a statistically significant reduction in chromosomal aberrations.[26]

7.2.14 Anti-asthmatic Activity

Kim et al., induced asthma in Balb/c mice by ovalbumin sensitization. Piperine (4.5 and 2.25 mg/kg) was orally administered 5 times a week for 8 weeks and was found that piperine-treated groups had suppressed eosinophil infiltration, allergic airway inflammation and airway hyper-responsiveness, and these occurred by suppression of the production of interleukin-4, interleukin-5, immunoglobulin E and histamine.[27]

7.2.15 Toxicity Studies

Dogra et al., (2004) administered piperine (1.12, 2.25, and 4.5 mg/kg, PO) for 5 days consecutively to determine the immunotoxicity in Swiss male mice. They noted that piperine at 2.25 and 4.5 mg/kg caused a significant reduction in total leukocyte counts, increase in the percentage of neutrophils and suppressed the mitogenic response of B-lymphocyte to lipopolysaccharide. Treatment at the highest dose, however, resulted in significant decrease in the weight of the spleen, thymus and mesenteric lymph nodes, but not of peripheral lymph nodes. Since piperine at dose 1.12 mg/kg had no immunotoxic effect, it may be considered as immunologically safe "no observed adverse effect level dose".[28]

7.2.16 Miscellaneous Activities

A. Li et al., evaluated the inhibitory effects of piperine on caries-related bacteria and glucan on dental plaque *in vitro*. They concluded that piperine had greater than 40 percent inhibitory effect on soluble glucan synthesis. Both insoluble and soluble glucan synthesis were inhibited by piperine.[29]

B. Veerareddy et al., prepared different formulations (lipid nanospheres of piperine, lipid nanosphere of piperine with stearylamine, pegylated lipid nanospheres of piperine and evaluated them as antileishmanial in BALB/c mice infected with Leishmania donovani AG83, for 60 days. A single dose (5 mg/kg, i.v.) of piperine and different formulation were injected. Mice were sacrificed after 15 days of treatment with piperine or formulations and Leishmania donovani Unit was counted. They concluded that piperine reduced the parasite burden in liver and spleen by 38 and 31 percent after 15-day post infection, respectively. Lipid nanospheres of piperine reduced the parasite burden in liver and spleen by 63 and 52 percent respectively. Pegylated lipid nanospheres of piperine reduced the parasite burden in liver and spleen by 78 and 75 percent respectively. Lipid nanosphere of piperine with stearylamine reduced the parasite burden in liver and spleen by 90 and 85 percent respectively as compared to the control.[30]

C. Volak et al., evaluated that piperine was a relatively selective noncom-petitive inhibitor of CYP3A with IC50 = 5.5 +/– 0.7 μM, K(i) = 5.4 +/– 0.3 μM.[31] Ononiwu et al., evaluated that piperine produced a dose dependent (at 20 mg/kg, 22.2 percent and 142 mg/kg, 334.6 percent) increase in gastric secreation in albino rats which may be due to stimulation of H2 receptors.[32]

REFERENCES

1. Singh, YN. Kava an overview. *J Ethnopharmacol* 1992; 37: 18-45.
2. Song Li, Che Wang, Minwei Wang, Wei Li, Kinzo Matsumoto, Yiyuan Tang. Antidepressant like effects of piperine in chronic mild stress treated mice and its possible mechanisms. *Life Sci* 2007; 80: 1373-1381.
3. Wattanathorn G. Antidepressant activity of piperine in animal studies. *Pharmacology* 1999; 24: 173-177.
4. Kulkarni SK, Bhutani MK, Bishnoi M. Antidepressant activity of curcumin: involvement of serotonin and dopamine system. *Psychopharmacology* (Berl) 2008; 201: 435-442.
5. Zhao JQ, Du GZ, Xiong YC, *et al.* Attenuation of beryllium induced hepatorenal dysfunction and oxidative stress in rodents by combined effect of gallic acid and piperine. *Arch Pharm Res* 2007; 30: 1575-1583.
6. Nirala SK, Bhadauria M, Mathur R, *et al.* Influence of alpha-tocopherol, propolis and piperine on therapeutic potential of tiferron against beryllium induced toxic manifestations. *J Appl Toxicol* 2008; 28: 44-54.
7. Vladimir B, Muhammed M, Edward PN. Piperine, An alkaloid derived from black pepper increases serum response of beta-carotene during 14 days of oral beta-carotene supplementation. *Nut Res* 1999; 19: 381-388.
8. Vijayakumar RS, Surya D, Nalini N. Antioxidant efficacy of black pepper (*Piper nigrum* L.) and piperine in rats with high fat diet induced oxidative stress. *Redox Rep* 2004; 9: 105-110.
9. Vladimir B, Muhammed M, Prakash L. Piperine derived from black pepper increases the plasma levels of coenzyme Q10 following oral supplementation. *J Nut Biochem* 2000; 11: 109-113.
10. Pattanaik S, Hota D, Prabhakar P, *et al.* Pharmacokinetic interaction of single dose of piperine with steady-state carbamazepine in epilepsy patients. *Phytother Res* 2009; 12: 34-37.
11. Kasibhatta R, Naidu MU. Influence of piperine on the pharmacokinetics of nevirapine under fasting conditions: a randomized, crossover, placebo-controlled study. *Drugs R D* 2007; 8: 383-391.

12. Durgaprasad S, Pai CG, Vasanthkumar Alvres JF, *et al.* A pilot study of the antioxidant effect of curcumin in tropical pancreatitis. *Indian J Med Res* 2005; 122: 315-318.
13. Choi BM, Kim SM, Park TK, *et al.* Piperine protects cisplatin-induced apoptosis via heme oxygenase-1 induction in auditory cells. *J Nut Biochem* 2007; 18: 615-622.
14. Selvendiran K, Padmavathi R, Magesh V, *et al.* Preliminary study on inhibition of genotoxicity by piperine in mice. *Fitoterapia* 2005a; 76: 296-300.
15. Pathak N, Khandelwal S. Cytoprotective and immunomodulating properties of piperine on murine splenocytes: An *in vitro* study. *Eur J Pharmacol* 2007; 576: 160-170.
16. Park BS, Son DJ, Park YH, *et al.* Antiplatelet effects of acidamides isolated from the fruits of *Piper longum* L. *Phytomedicine* 2007; 14: 853-855.
17. Kumar S. Piperine inhibits TNF-α induced adhesion of neutrophils to endothelial monolayer through suppression of NF-κB and IκB kinase activation. *J Pharmacol* 2007; 575: 177-186.
18. Pradeep CR, Kuttan G. Piperine is a potent inhibitor of nuclear factor-kappaB (NF-kappaB), c-Fos, CREB, ATF-2 and proinflammatory cytokine gene expression in B16F-10 melanoma cells. *Int Immunopharmacol* 2004; 4: 1795-1803.
19. Taqvi SI, Shah AJ, Gilani AH. Blood pressure lowering and effects of piperine. *J Cardiovasc Pharmacol* 2008; 52: 452-458.
20. Matsuda H, Ninomiya K, Morikawa T, *et al.* Protective effects of amide constituents from the fruit of Piper chaba on D-galactosamine/TNF-alpha-induced cell death in mouse hepatocytes. *Bioorg Med Chem Lett* 2008; 18: 2038-2042.
21. Vijayakumar RS, Nalini N. Piperine, an active principle from Piper nigrum, modulates hormonal and apo lipoprotein profiles in hyperlipidemic rats. *J Basic Clin Physiol Pharmacol* 2006; 17: 71-86.
22. Panda S, Kar A. Piperine lowers the serum concentrations of thyroid hormones, glucose and hepatic 5′D activity in adult male mice. *Horm Metab Res* 2003; 35: 523-526.
23. Piyachaturawat P, Pholpramool C. Enhancement of fertilization by Piperine in Hamsters. *Cell Biol Int* 1997; 21: 405-409.
24. Manoharan S, Balakrishnan S, Menon VP, *et al.* Chemopreventive efficacy of curcumin and piperine during 7,12-dimethylbenz[a]anthracene-induced hamster buccal pouch carcinogenesis. *Singapore Med J* 2009; 50: 139-146.
25. Duessel S, Heuertz RM, Ezekiel UR. Growth inhibition of human colon cancer cells by pants compounds. *Clin Lab Sci* 2008; 21: 151-157.
26. Wongpa S, Himakoun L, Soontornchai S, *et al.* Antimutagenic effects of piperine on cyclophosphamde-induced chromosome aberrations in rat bone marrow cells. *Asian Pac J Cancer Prev* 2007; 8: 623-627.

27. Kim SH, Lee YC. Piperine inhibits eosinophil infiltration and airway hyperresponsiveness by suppressing T-cell activity and Th2 cytokine production in the ovalbumin-induced asthma model. *J Pharm Pharmacol* 2009; 61: 353-359.
28. Dogra RK, Khanna S, Shanker R. Immunotoxicological effects of piperine in mice. *Toxicology* 2004; 196: 229-236.
29. Li M, Liu Z. *In vitro* effect of Chinese herb extracts on caries-related bacteria and glucan. *J Vet Dent* 2008; 25: 236-239.
30. Veerareddy PR, Vobalaboina V, Nahid A. Formulation and evaluation of oil-in-water emulsions of piperine in visceral leishmaniasis. *Pharmazie* 2004; 59: 194-197.
31. Volak LP, Ghirmai S, Cashman JR, *et al.* Curcuminoids inhibit multiple human cytochromes P450, UDP-glucuronosyltransferase, and sulfotransferase enzymes, whereas piperine is a relatively selective CYP3A4 inhibitor. *Drug Metab Dispos* 2008; 36: 1594-1605.
32. Ononiwu IM, Ibeneme CE, Ebong OO. Effects of piperine on gastric acid secretion in albino rats. *Afr J Med Sci* 2002; 31: 293-295.

Chapter 8

Pharmacological Profile of Oxoaporphine Alkaloid, Liriodenine

8.1 INTRODUCTION

Liriodenine (Fig. 8.1) is an oxoaporphine alkaloid obtained from *Annona cherimolia, Cananga odorata, Fissistigma glaucescens, Liriodendron tulipifera, Michelia compressa, Pseuduvaria setosa* and *Polyalthia sclerophylla*.[1,2]

Liriodenine is a relatively lesser-known alkaloid, but during the recent past has attracted the attention of the scientific community as diverse pharmacological activities have been reported. In this chapter, we have documented the pharmacological investigations carried out on the alkaloid.

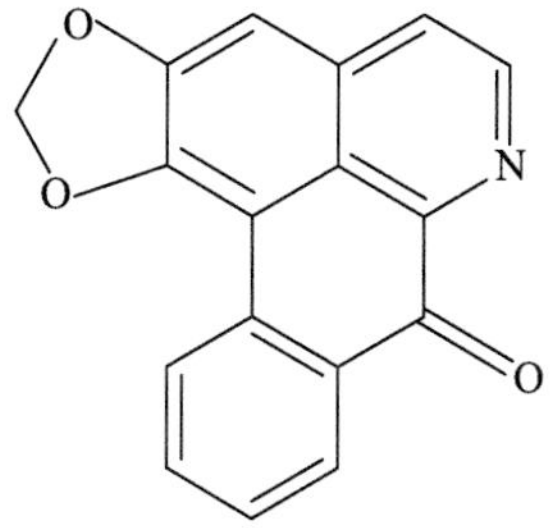

Figure 8.1 Structure of Liriodenine.

8.2 PHARMACOLOGY

8.2.1 Anticancer

Liriodenine, isolated from *Cananga odorata*, was found to be a potent inhibitor of topoisomerase II (EC 5.99.1.3) both *in vivo* and *in vitro*. Liriodenine treatment of SV40 (simian virus 40)-infected CV-1 cells caused highly catenated SV40 daughter chromosomes, a signature of topoisomerase II inhibition. Strong catalytic inhibition of topoisomerase II by liriodenine was confirmed by *in vitro* assays with purified human topoisomerase II and kinetoplast DNA. Liriodenine also caused low-level protein-DNA cross-links to pulse-labeled SV40 chromosomes *in vivo*, suggesting that it may be a weak topoisomerase II poison. Verapamil did not increase either liriodenine-induced protein-DNA cross-links or catalytic inhibition of topoisomerase II in SV40-infected cells. This indicates that liriodenine is not a substrate for the verapamil-sensitive drug efflux pump in CV-1 cells.[3]

In a study, the effect of liriodenine on the growth and viability of human lung cancer cells was investigated. Liriodenine suppresses proliferation of A549 human lung adenocarcinoma cells in a dose- and time-dependent manner. Flow cytometric analysis showed that liriodenine blocked cell cycle progression at the G2-M phase. Induction of G2-M arrest by liriodenine was accompanied by reduction of G1 cyclin (D1) and accumulation of G2 cyclin (B1). *In vitro* kinase activity assay demonstrated that the enzymatic activity of the cyclin B1/cyclin-dependent kinase 1 complex was reduced in liriodenine-treated cells. More importantly, incubation with liriodenine led to activation of caspases and apoptosis in A549 cells. The results indicate that liriodenine exerts potent anti-proliferative and apoptosis-inducing effects on human lung cancer cells.[4]

In yet another study, liriodenine, isolated from the leaves of *Michelia compressa* was investigated for its impact to arrest the cell cycle, production of nitric oxide and p53 expression in human hepatoma cell lines, including wild-type p53 (Hep G2 and SK-Hep-1). As evidenced by flowcytometric studies, liriodenine induced cell cycle G(1) arrest and inhibited DNA synthesis in Hep G2 and SK-Hep-1 cell lines. The p53, iNOS expression and intracellular NO level were markedly increased in Hep G2 cells after liriodenine treatment. These results demonstrate that nitric oxide production and p53 expression are critical factors in liriodenine-induced growth inhibition in human wild-type p53 hepatoma cells.[5]

8.2.2 Antimicrobial

8.2.2.1 Antinematode

Liriodenine methiodide prepared from liriodenine inhibited rat *P. carinii* in a culture screen using human embryonic lung cells (HEL) sheeted in 24 well tissue culture plates. Both compounds at 1.0 μg/ml reduced *P. carinii* during the 7 days evaluation period so that on day 7 numbers of trophic forms were 20 percent or less of untreated control cultures and comparable to trimethoprim/sulfamethoxazole-treated cultures (50/250 μg/ml).[6]

8.2.2.2 Antimalarial

Liriodenine, isolated from the aerial part of *Pseuduvaria setosa,* displayed antituberculosis activity against *Mycobacterium tuberculosis,* being most active at minimum inhibitory concentration (MIC) of 12.5 μg/ml. It also demonstrated antimalarial activity against Plasmodium falciparum with the 50 percent inhibitory concentration (IC50) of 2.8 μg/ml. Furthermore, the alkaloid was strongly cytotoxic to both epidermoid carcinoma (KB) and breast cancer (BC) cell lines.[7]

8.2.3 Cardiovascular System

8.2.3.1 Vasodilator activity

The effect of liriodenine and norushinsunine, isolated from *Annona cherimolia*, were studied in the rat aorta in order to examine their mechanism of action. Both alkaloids (10^{-7}–10^{-4} mol/l) showed relaxant effects on the contractions elicited by 10^{-6} mol/l noradrenaline (NA) or 80 mmol/l KC1, however while liriodenine showed a nonspecific relaxant action on both spasmogens, norushinsunine was more potent on KC1-induced contraction. In Ca2+ -free medium, both alkaloids (0.1 mmol/l) inhibited the responses elicited by NA, but not those elicited by caffeine. This inhibitory action occurred when the alkaloids were present during the release of the Ca2+ internal stores or during the refilling process.[8]

8.2.3.2 Anti-arrhythmic activity

This study investigated whether anti-arrhythmic dose of 10^{-7} g/kg had effects on the left ventricular (LV)-arterial coupling in Wistar rats. LV pressure and ascending aortic flow signals were recorded to construct the ventricular and arterial end-systolic pressure-stroke volume relationships to calculate LV end-systolic elastance (Ees) and effective arterial volume elastance (Ea), respectively. The optimal afterload (Q_{load}) determined by the ratio of Ea to Ees was used to measure the optimality of energy transmission from the left ventricle to the arterial system. Liriodenine at the dose of 10^{-6} g/kg produced a significant fall of 2.0 percent in HR and a significant rise of 5.8 percent in CO, but no significant change in Pes. Moreover, liriodenine administration of 10^{-6} g/kg to rats significantly decreased Ees by 8.5 percent and Ea by 10.6 percent , but did not change Q_{load}. It was concluded that liriodenine at the dose of 10^{-7} g/kg has no effects on the mechanical properties of the heart, the vasculature and the matching condition for the left ventricle coupled to its vasculature in rats.[9]

8.2.3.3 Cardio protective

The present study evaluated the protection conveyed by liriodenine to myocardium and coronary endothelial cells under conditions of ischemia–reperfusion and to assess the involvement of a nitric oxide (NO)-dependent mechanism. In the Langendorff model utilizing Sprague–Dawley rat hearts, the left main coronary artery was occluded for 30 min and reperfusion for 120 min. Liriodenine (1 μM) significantly promoted the recovery of coronary flow and decreased myocardial infarction compared with vehicle-treated hearts. The drug attenuated the reduction of endothelial reactivity and NO release. Liriodenine prevented eNOS reduction in serum-deprived HUVEC and ischemia–reperfusion hearts. The vascular and cardio-protective effects were reversed by N^G-nitro-L-arginine methyl ester. Another Na^+ and K^+ channel blocker with similar activities as liriodenine (quinidine) failed to protect endothelial cells and myocytes.[10]

8.2.4 Central Nervous System (C.N.S)

8.2.4.1 Muscarinic receptor antagonist

Liriodenine was found to be a muscarinic receptor antagonist in guinea-pig trachea as revealed by its competitive antagonism of carbachol (pA2 = 6.22 +/− 0.08)-induced smooth muscle contraction. It was slightly more potent than methoctramine (pA2 = 5.92 +/− 0.05), but was less potent than atropine (pA2 = 8.93 +/− 0.07), pirenzepine (pA2 = 7.02 +/− 0.09) and 4-diphenylacetoxy-N-methylpiperidine (4-DAMP, pA2 = 8.72 +/− 0.07). 3. Liriodenine was also a muscarinic antagonist in guinea-pig ileum (pA2 = 6.36 +/− 0.10) with a pA2 value that closely resembled that obtained in the trachea. High concentration of liriodenine (300 μM) partially depressed the contractions induced by U-46619, histamine, prostaglandin F2 alpha, neurokinin A, leukotriene C4 and high K^+ in the guinea-pig trachea.[11]

8.2.4.2 Dopaminergic

In this study, the inhibitory effects of liriodenine, on dopamine biosynthesis and L-DOPA-induced dopamine content increases in PC12 cells were investigated. Treatment of PC12 cells with 5-10 μM liriodenine significantly decreased the intracellular dopamine content in a concentration-dependent manner (IC50 value, 8.4 μM). Liriodenine was not cytotoxic toward PC12 cells at concentrations up to 20 μM. Tyrosine hydroxylase and aromatic L-amino acid decarboxylase activities were inhibited by 10 μM liriodenine to 20-70 and 10-14 percent of control levels at 3-12 hr, respectively; Tyrosine hydroxylase activity was more influenced than L-amino acid decarboxylase activity. In addition, 10 μM liriodenine reduced L-DOPA (20-100 μM)-induced increases in dopamine content.[12]

REFERENCES

1. Cordell G. *Introduction to Alkaloids: A Biogenetic Approach.* Wiley and Sons, New York, 1981.
2. Said M, Hadi A, Awang K. The alkaloids of *Polyalthia sclerophylla.* Seminar Peneyelidikan Jangka, Pendek (Vot F). 2003.
3. Woo SH, Reynolds MC, Sun NJ, *et al.* Inhibition of topoisomerase II by liriodenine. *Biochem Pharmacol* 1997; 54: 467-473.

4. Chang HC, Chang FR, Wu YC, *et al.* Anti-cancer effect of liriodenine on human lung cancer cells. *The Kaohsiung J of Med Sci* 2004; 20: 365-371.
5. Hsieh TJ, Liu TZ, Chern CL, *et al.* Liriodenine inhibits the proliferation of human hepatoma cell lines by blocking cell cycle progression and nitric oxide-mediated activation of p53 expression. *Food Chem Toxicol* 2005; 43: 1117-1126.
6. Bartlett MS, Queener SF, Clark AM, *et al. Derivatives of natural products, liriodenine methiodide and axisonitrile-3, are effective against pneumocystis carinii in vitro.* Program Abstr First Natl Conf Hum Retrovir Relat Infect Natl Conf Hum Retrovir Relat Infect 1993; 16: 97.
7. Wirasathien L, Boonarkart C, Suttisr P. Biological Activities of Alkaloids from *Pseuduvaria setosa. Pharm Biol* 2006; 44: 274-278.
8. Chuliá S, Noguera MA, Ivorm MD, *et al.* Vasodilator Effects of Liriodenine and Norushinsunine, Two Aporphine Alkaloids Isolated from *Annona cherimolia* in Rat Aorta. *Pharmacology* 1995; 50: 380-387.
9. Chang KC, Su MJ, Peng YI, Shao CC, Wu YC, Tseng YZ. Mechanical effects of liriodenine on the left ventricular-arterial coupling in Wistar rats: pressure-stroke volume analysis. *Br J Pharmacol* 2001; 133(1): 29-36.
10. Luen CW, Hu CC, Yang-Chang WU, *et al.* The vascular and cardio protective effects of liriodenine in ischemia–reperfusion injury via NO-dependent pathway. *Nitric oxide* 2004; 11: 307-315.
11. Lin CH, Chang GJ, Su MJ, *et al.* Pharmacological characteristics of liriodenine, isolated from *Fissistigma glaucescens,* a novel muscarinic receptor antagonist in guinea-pigs. *Br J Pharmacol* 1994; 113: 275-281.
12. Mei JC, Lee JJ, Jung YY, *et al.* Liriodenine inhibits dopamine biosynthesis and L-DOPA-induced dopamine content in PC12 cells. *Arch Pharmacol Res* 2007; 30: 984-90.

Chapter 9

Recent Experimental Advances in Hepatoprotective Potential of Andrographolide

9.1 INTRODUCTION

Andrographis paniculata Nees. (Acanthaceae) is found throughout India. In Ayurveda, it is commonly known as *Bhunimba. A. paniculata* is an annual herb. Branches are quadrangular and dark green in color. The leaves are opposite each other, shortly staked, green above, pale below and pinnate. The flowers are small and whitish-purple in color. The fruit is a capsule having many seeds. The root is unbranched, contains many rootlets and has a gray color.[1]

In Ayurveda, *A. paniculata* is used in the treatment of jaundice, hepatomegaly, constipation and malaria.[2-4] The whole plant is used in medicine and *Liquid Extract Kalmegh* is the official preparation.[5,6]

9.2 DITERPENE LACTONES OF *A. PANICULATA*

The herb is reported to contain diterpene lactones including, andrographolide (Fig. 9.1), neoandrographolide, 14-deoxy-11-oxo andrographolide, 14-deoxy-11, 12-oxo andrographolide, 14-deoxy andrographolide, echiodin, kalmeghin, homandrographolide, andrographane and andrographone.[7,8] According to official standards, the drug should not contain andrographolide of not less than 1 percent.[9,10]

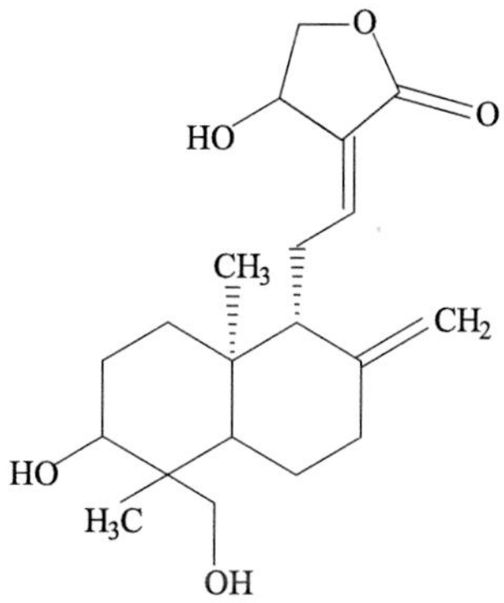

Figure 9.1 Structure of andrographolide.

9.3 PREVIOUS REPORTED WORK ON ANDROGRAPHOLIDE

A clinical study by Chaturvedi et al., reported efficacy of *A. paniculata* in infective hepatitis.[11] Choudhury and Poddar reported *in vivo* and *in vitro* effect of andrographolide and *A. paniculata* extract on hepatic lipid peroxidation.[12] Handa and Sharma reported hepatoprotective activity of andrographolide against carbon tetrachloride induced hepatotoxicity.[13] Rana and Avadhoot studied the hepatoprotective effect of crude *A. paniculata* against carbon tetrachloride-induced liver damage.[14] Exploring the hepatoprotective mechanism of andrographolide, Shukla et al., reported choloretic effect of the same in rats and guinea pigs.[15]

In a study, Visen et al., observed the protective action of andrographolide in rat hepatocytes against paracetamol-induced damage.[16] An investigatory study by Trivedi and Rawal observed hepatoprotective and anti-oxidant activity of alcoholic extract of *A. paniculata* against BHC induced hepatotoxicity in rats.[17] Kapil et al., reported antihepatotoxic effects of andrographolide, andrographiside and neoandrographolide on hepatotoxicity induced in mice by carbon tetrachloride or tert-butylhydroperoxide intoxication.[18]

9.4 RECENT FINDINGS

Visen et al., studied the possible hepatoprotective effect of andrographolide against galactosamine induced hepatotoxicity. Andrographolide showed concentration dependent (1-100 μg/ml) activity against galactosamine (400 μg/ml) induced acute injury in isolated monkey hepatocytes, following 48 hr of the incubation at 370 C. Andrographolide significantly increased the percent viability (trypan blue exclusion and oxygen uptake tests) of cells and was capable of preserving 90-100 percent cell integrity. The cellular leakage of enzymes and bile contents following incubation with galactosamine was significantly modulated by andrographolide treatment.[19]

Singha et al., explored the active principles of *A. paniculata* against ethanol-induced toxicity in mice. The mice are pretreated intra-peritoneally with different doses (62.5, 125, 250, and 500 mg/kg) of body weight of mice of andrographolide and arabinogalactan proteins, isolated from *A. paniculata,* for 7 days and then ethanol (7.5 g/kg of body weight) was injected, IP Pretreatment of mice with andrographolide and arabinogalactan proteins at 500 mg/kg of body weight and 125 mg/kg of body weight respectively, minimized the toxicity in comparison to ethanol treated group and the results were comparable with silymarin.[20]

Trivedi and coworkers investigated the effect of andrographolide on the hepatocellular anti-oxidant defense system and lipid peroxidation of controlled mice, mice treated with hexachlorocyclohexane only, and andrographolide + hexachlorocyclohexane. The activities of glutathione, glutathione reductase, glutathione peroxidase, superoxide dismutase, and catalase showed significant (P < or = 0.05) increases, while γ-glutamyl transpeptidase and glutathione-s-transferase showed significant decreases (P < or = 0.05) in andrographolide-supplemented mice as compared with BHC-treated mice.[21]

REFERENCES

1. Nadkarni KM. *Indian Materia Medica.* 3rd ed. Bombay, India: Popular Book Depot, 1992; p. 76.
2. Kirtikar KR, Basu BD. *Indian Medicinal Plants.* In: Basu LM, (Ed). Reprinted edition Drona Publications, Allahabad, India: 1989; p. 27.
3. Singh AP. *Dravyaguna Vijnana.* New Delhi, India: Chaukhambha Orientalia, 2005; p. 321.
4. Singh AP, Malhotra S, Munjal G. Hepatoprotective Potential of Phytomedicines. *Gastroenterol Today* 2002; 6: 76-77.
5. *Database on medicinal plants used in Ayurveda.* Vol. 5. New Delhi, India: Central Council for Research in Ayurveda and Siddha Govt. of India: New Delhi, 2002; p. 269.
6. Singh AP. *A. paniculata*-A Valuable Medicinal Plant. *Pharmabuzz* 2006; 1: 36-39.
7. Singh AP. *Materia Medica and Herbal Pharmacology.* New Delhi, India: Chaukhambha Orientalia, 2005; p. 178.
8. Singh AP. *Glossary of Medicinal Plants used in Ayurveda***.** Ist ed. Jodhpur, India: Scientific Publishers, 2006; pp. 56.
9. Anonymous. *Pharmacopoeial standards of Ayurvedic formulations.* Central Council for Research in Ayurveda and Siddha Govt. of India: New Delhi, 1987; p. 137.
10. Chakravarti RN, Chakravarti D. Andrographolide, the active constituent of Andrographis paniculata Nees; a preliminary communication. *Ind Med Gaz* 1951; 86: 96-97.

11. Chaturvedi GN, Tomar GS, Tiwari SK, *et al.* Clinical studies on kalmegh *(A. paniculata)* in infective hepatitis. *J Int Inst Ayur* 1983; 2: 208-211.
12. Choudhury BR, Poddar MK. Andrographolide and kalmegh (*A. paniculata*) extract: *in vivo* and *in vitro* effect on hepatic lipid peroxidation. *Meth Find Exp Clin Pharmacol* 1984; 6: 481-485.
13. Handa SS, Sharma A. Hepatoprotective activity of andrographolide from *A.paniculata* against carbon tetrachloride. *Indian J Med Res* 1990; 92276-9283.
14. Rana AC, Avadhoot Y. Hepatoprotective effects of *A. paniculata* against carbon tetrachloride-induced liver damage. *Pharmacol Res* 1991; 14: 93-95.
15. Shukla B, Visen PK, Patnaik GK, *et al.* Cholretic effect of Andrographolide in rats and guinea pigs. *Planta Med* 1992; 58: 146-149.
16. Trivedi NP, Rawal UM. Hepatoprotective and antioxidant property of *A. paniculata* (Nees) in BHC induced liver damage in mice. *Indian J Exp Biol* 2001; 39: 41-46.
17. Visen PK, Shukla B, Patnaik GK, *et al.* Andrographolide protects rat hepatocytes against paracetamol-induced damage. *J Ethnopharmacol* 1993; 40: 131-136.
18. Kapli A, Koul IB, Banerjee SK, *et al.* Antihepatotoxic effects of major diterpenoid constituents of *A. paniculata. Biochem Pharmacol* 1993; 46: 182-185.
19. Pradeep KS Visen, Saraswat B, Vuksan V, Dhawan BN. Effect of Andrographolide on Monkey Hepatocytes against Galactosamine Induced Cell Toxicity: An *In-Vitro* Study. *J Comp Integ Med* 2007; 4: 125-129.
20. Singh PK, Roy S, Dey S. Protective activity of andrographolide and arabinogalactan proteins from *A. paniculata* Nees. against ethanol-induced toxicity in mice. *J Ethnopharmacol* 2007; 111: 13-21.
21. Trivedi NP, Rawal UM, Patel BP. Hepatoprotective effect of andrographolide against hexachlorocyclohexane-induced oxidative injury. *Integ Can Ther* 2007; 6: 271-280.

Chapter 10

Profile of Hypericin-Napthodianthrone from *Hypericum perforatum* Linn.

10.1 NAPTHODIANTHRONES OF *HYPERICUM PERFORATUM* LINN.

Napthodianthrones are derivatives of anthracene. The napthodianthrones of *Hypercium perforatum* includes hypericin (Fig. 10.1), pseudohypericin and isohypericin. The fresh plant contains protohypericin and protopseudohypericin and in presence of light, are converted into hypericin and pseudohypericin.[1]

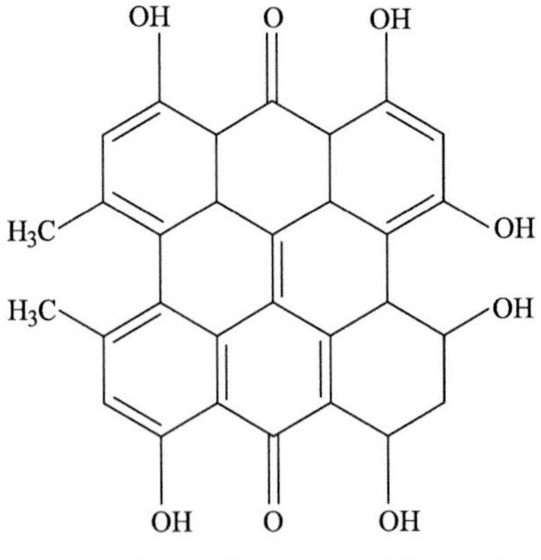

Figure 10.1 Structure of hypericin.

Accroding to Newallet et al., the total amount of napthodianthrones ranges from 0.05-0.15 percent.[1] According to Upton, the hypericin content of dried flowers can be up to 1.8 percent. *H. perforatum* extract used in most of the clinical trials has been standardized to contain at least 0.3 percent hypericin.[2] However, hypericin is now days regarded as a marker compound in *Hypericum perforatum* extracts.[3]

10.2 BIOSYNTHESIS OF HYPERICIN *IN HYPERICUM ELODES*

Piovan et al., investigated sepals of *Hypericum elodes* with electro spray ionization mass spectrometry (ESI-MS/MS). The sepals contain red glands (stipitate glands with red-colored heads). Piovan states that the occurrence of hypericin in red glands of *Hypericum elodes* strongly suggests the taxonomic position of the section Elodes within the genus Hypericum. He further states that the ability to carry out the biosynthetic pathway leading to the naphthodianthrone compounds, rather than the absolute amounts produced, should be regarded as a chemical marker of the phylogenetically more advanced sections of genus Hypericum.[4]

10.3 ANTIDEPRESSANT ACTIVITY

10.3.1 Interaction with Cholinergic and σ Receptors

Raffa showed that hypericin (1.0 μM) has modest affinity for muscarinic cholinergic receptors (49 percent) as well as σ receptors (48 percent). σ receptors are located in areas of the brain involved with the regulation of emotions, such as the limbic system, and a role for σ receptors in the antidepressive action of synthetic antidepressants has been postulated.[5]

10.3.2 Hypericin and Monoamine Oxidase Inhibition

When tested in laboratory conditions, a virtually irreversible inhibition of monoamine oxidase (MAO) type A and type B was reported in rat brain mitochondria exposed to hypericin.[6] Unable to reproduce the inhibition with hypericin, researchers did obtain MOAI with a Hypericum extract.[7] Tests done on hypericin (not the extract) found that it did not exhibit monoamine oxidase inhibition *in vitro*.[8]

10.4 ANTICANCER ACTIVITY

10.4.1 MX-1 Human Mammary Carcinoma

Thomas examined the potential of hypericin as an agent for chemotherapy and for photodynamic therapy of the neoplastic disease. The model systems used to test for activity were a mitochondrial succinoxidase enzyme activity assay, an EMT6 mouse mammary carcinoma clonogenic assay, and an MX-1 human mammary carcinoma antitumor assay. Parallel experiments were conducted in the presence and absence of light. Pure hypericin and acetone extracts of *H. perforatum* were tested for activity and yielded similar results. In the dark, hypericin inhibited succinoxidase activity in isolated mitochondrial membranes, but had no effect on EMT6 cells or MX-1 tumors. However, photoactivated hypericin exhibited activity in all test systems.

Photoactivated hypericin inhibited succinoxidase activity at μM/mg mitochondrial protein levels, EMT6 cell clonogenicity at μM levels, and MX-1 relative tumor weight growth in IP and animals were dosed at 1 mg/kg body weight levels. Experiments designed to elucidate the mechanistic parameters of hypericin's biological activity were also conducted. The results show that photosensitized inhibition of mitochondrial succinoxidase was drug dose-, light dose- and wavelength-dependent with the greatest inhibition occurring by 600 nm light activation. The photodynamic effect of hypericin on EMT6 cells was examined under hypoxic and aerobic conditions, and photocytotoxicity was found to be oxygen-dependent.

These findings suggest that type I oxygen-independent photo-oxidation reactions do not contribute significantly to hypericin's photodynamic mechanism of action. Fluorescence photomicrographs of hypericin in EMT6 cells show that hypericin is distributed throughout the plasma and cytosolic membranes but is absent from the nucleus, indicating that the events which lead to photocytotoxicity may not involve direct damage to nuclear DNA.

10.4.2 Glioma

Couldwell et al., investigated glioma-inhibitory activity of hypericin. Protein kinase C has been implicated in the formation and proliferation of tumors. Hypericin inhibited the growth of glioma cell lines *in vitro* and induced glioma cell death due to inhibition of Protein kinase C, as measured by [3H]-thymidine uptake. The glioma-inhibitory activity of hypericin was comparable to or greater than tamoxifen (IC50 $<$ 10 versus 10 μM/L, respectively). The activity was enhanced by approximately 13 percent from exposure of hypericin to visible light.[9]

10.4.3 *In vitro* Cytotoxic and *in vivo* Antitumoral Activity

Vandenbogaerde investigated *in vitro* cytotoxic and *in vivo* antitumoral properties of photosensitized hypericin. *In vitro* studies demonstrated a dramatic difference in sensitivity of cell lines towards photo-activated hypericin using a neutral red assay. Using confocal laser microscopy it was shown that the cytotoxicity correlated with the cellular uptake of hypericin. The *in vivo* antitumoral activity hypericin was tested on A431 cell xenografts in athymic nude mice. A dose-dependent antitumoral effect was observed by intraperitoneal administration of hypericin in light-treated animals; a complete inhibition of the tumor growth was achieved from 50 μg (2.5 mg/kg) hypericin IP administered. The photosensitizer accumulated in A431 xenografts after local light irradiation. These results showed hypericin is a potential photochemotherapeutic agent.[10]

10.4.4 Clinical Study on Skin Cancer, Basal Cell Carcinoma and Squamous Cell Carcinoma

Alecu et al., examined the topical use of hypericin as a photodynamic anti-cancer therapy in 19 patients with skin cancer—11 patients (median age 66) diagnosed with basal cell carcinoma and eight patients (ages 39-72) with squamous cell carcinoma, five of whom had not received prior treatments for their cancer. Treatment consisted of intralesional injections of hypericin followed with irradiation using visible light to photo-activate hypericin.

For ethical reasons, surgery was used after the experimental therapy to remove residual tumor lesions and to make an assessment of possible impairments caused by the phototherapy. Among the squamous cell carcinoma patients, tumors were reduced by 33 percent after treatment with hypericin (40-100 μg hypericin i.l. 3-5 μ/week for 2-6 weeks) and in one case the treatment appeared to cause complete remission.

One female patient showed no tumor reduction after 10 injections (540 μg hypericin) over 15 days into a tumor of the lower lip (30/20 μ 10 mm in size). Analysis of the results showed that tumor reduction was dependent on initial tumor size and the dosage of hypericin administered. The analysis also revealed that in total 1,000 μg plus would be required in order for hypericin to reach acceptability as an alternative treatment for squamous cell carcinoma.

Among the patients with basal cell carcinoma, treatment with hypericin (40-200 μg hypericin intralesionally, 3-5 μ/week for 2-6 weeks) resulted in tumor reductions of less than 10 percent to as much as 100 percent in two cases with complete remission. For all 11 patients, photodynamic treatment with hypericin represented their first therapy for cancer. All these patients refused follow-up surgery and 5 months later showed no sign of the cancer remaining or recurring. In total, these patients received from 1,500-3,000 μg of hypericin. The only side-effect observed was edema and erythema of a mild transient nature at the tumor site, which occurred in three patients in the basal cell carcinoma group and two in the squamous cell group.[11]

10.5 ANTIVIRAL ACTIVITY

Takahashi et al., while investigating antiretroviral activity of hypericin and pseudohypericin found that both specifically inhibited protein kinase C with IC50 values 1.7 μg/ml and 15 μg/ml, respectively, and show antiproliferative activity against mammalian cells. The data suggested that antiretroviral activity of hypericin and pseudohypericin could be attributable to the inhibition of some phosphorylation involved by protein kinase C during viral infection of cells.[12]

10.5.1 Clinical Studies on Antiviral Activity of Hypericin

A. Mcauliffe et al., described a phase I study of intravenous dosed synthetic hypericin. HIV-infected patients with less than or equal to 300 CD4 cells were eligible for enrollment. Patients received intravenous hypericin at 0.25 or 0.5 mg/kg BIW or 0.25 mg/kg thrice a day; with assessments of toxicity, pharmacokinetic and antiviral activity. Four patients received a single oral dose. Twenty five patients received hypericin from 1 to 24 weeks. Phototoxicity developed in all patients, varied in severity, and was dose limiting at 0.5 mg/kg twice a day.

Pharmacokinetic showed an area under the curve of 26.0 +/− 5.0 (Mean +/− SD) and 53.7 +/− 27.2 +/− μg/hr/ml at intravenous doses of 0.25 and 0.5 mg/kg, respectively; peak levels were 4.2 +/− 1.1 and 7.7 +/− 2.0 μg/ml and fell below 1 μg/ml at 5 and 11 hr. Elimination half lives were 23.7 +/− 7.3 and 35.3 +/− 9 hours ($P > 0.1$). Oral bioavailability was 22.3 +/− 7.7 percent. A consistent change in antiviral endpoints was not seen with intermittent IV dosing. Intravenous dosing of hypericin yields dose-limiting cutaneous phototoxicity of variable severity. Pharmacokinetic data predict chronic oral dosing should achieve sustained blood levels in an anti-retroviral range.[13]

B. Vonsover et al., investigated the antiretroviral efficacy of long term therapy with plant derived hypericin on p24 antigen and on viral load in AIDS patients. Eighteen seropositive symptomatic individuals received monotherapy with plant derived hypericin [intravenous 1 × 2 ml weekly plus 6 × 2 hypericum perforatum tablets per day] for periods ranging between 48-72 months. Plasma viral RNA loads were monitored using Nucleic Acid Amplification Assay (NASBA, Organon-Teknika), and p24 antigen with an HIV-1 Antigen Base Dissociation Assay (Organon-Teknika); they were followed during 40 months of therapy.

Among the 18 participants, 33 percent had measurable, pretreatment concentrations of serum p24 antigen, one became positive after 24 months and twelve remained p24 antigen negative throughout the follow-up period. Four patients exhibited significant long term declines in p24 antigen levels; the levels remained unchanged in two others. HIV-1 RNA was detected in 83 percent of the patients with a mean of maximal RNA load of 5.33 and 5.00 log 10 copies/ml in p24 antigen-positive or negative patients respectively.

A substantial decline in viral load (from 5.0 to 4.23 log 10 copies/ml) was observed in most of the p24 antigen positive and negative individuals. In some patients an increase in virus load was observed throughout a three year treatment period without effect on the clinical conditions of viral CMV, herpes or EBV complications being encountered.[14]

C. Gulick et al., treated 30 HIV-positive patients with synthetic hypericin 0.25 or 0.5 mg/kg intravenous(i.v.) twice weekly or 0.25 mg/kg i.v. 3 µ/week, or 0.5 mg/kg/day PO for 8 weeks). Owing to phototoxic effects, only 14 patients completed the study. No significant change was found in CD4 cell counts, HIV p24 antigen levels, or HIV RNA copies, and out of 23 patients available for evaluation, 11 developed severe cutaneous phototoxicity, including all who received oral hypericin.[15]

10.6 IMMUNOSUPPRESSIVE ACTIVITY

Panossian et al., examined the effects of hypericin on the release of inflammatory mediators by human immune system cells. At low concentrations (IC50 4-8 M), hypericin significantly and dose dependently inhibited the release of arachidonic acid and, as a result, of leukotriene B4. The authors found that hypericin strongly inhibited interleukin-1 α, possibly by way of inhibiting protein kinase C. However, an extract of St. John's wort proved far stronger than hypericin at inhibiting the release of interleukin-1 α and leukotriene B4. The inhibition of the mediator, results in a quelling of the inflammatory response (immunosuppression), and that apparently besides hypericin, other active constituents are involved.[16]

10.7 PHARMACOKINETICS OF HYPERICIN

Stock and Holzl tested a radioactive isotope of hypericin in mice and humans. The maximum blood level was at 6 hr. The blood level at 3.5 hr was 0.2 percent of the dose and at 8 hr was 1.9 percent of the initial dose.[17]

Kerb et al., investigated single-dose and steady-state pharmacokinetics of hypericin and pseudohypericin in 13 healthy volunteers by administration of *Hy. perforatum* (extract LI 160). Oral administration of 250, 750, and 1,500 µg of hypericin and 526, 1,578, and 3,156 µg of pseudohypericin resulted in median peak levels in plasma (Cmax) of 1.3, 7.2, and 16.6 µg/l for H and 3.4, 12.1, and 29.7 µg/l for pseudohypericin, respectively. The Cmax and the area under the curve values for the lowest dose were disproportionally lower than those for the higher doses. A lag time of 1.9 hr for H was remarkably longer than the 0.4 hr lag time for pseudohypericin.

Median half-lives for absorption, distribution, and elimination were 0.6, 6.0, and 43.1 hr after 750 µg of hypericin and 1.3, 1.4, and 24.8 hr after 1,578 µg of pseudohypericin, respectively. A fourteen-day treatment with 250 µg of hypericin and 526 µg of pseudohypericin three times a day resulted in median steady-state trough levels of 7.9 µg/l for hypericin and 4.8 μg/l for pseudohypericin after 7 and 4 days, respectively; the corresponding Css, max levels were 8.8 and 8.5 μg/l, respectively.

Kinetic parameters after intravenous administration of Hypericum extract (115 and 38 μg for hypericin and pseudohypericin, respectively) in two subjects corresponded to those estimated after an oral dosage. Both hypericin and pseudohypericin were initially distributed into a central volume of 4.2 and 5.0 l, respectively. The mean distribution volumes at steady state were 19.7 l for hypericin and 39.3 l for pseudohypericin, and the mean total clearance rates were 9.2 ml/min for hypericin and 43.3 ml/min for pseudohypericin. The systemic availability of hypericin and pseudohypericin from LI 160 was roughly estimated to be 14 and 21 percent, respectively. Treatment with extract of *Hypericum perforatum*, even in high doses, was well tolerated.[18]

REFERENCES

1. Newall CA, Anderson LA, Phillipson JD. *Herbal Medicines: A Guide for Health Care Professionals.* London: The Pharmaceutical Press, 1996.
2. Upton R (Ed.). *St. John's Wort Monograph.* American Herbal Pharmacopoeia, 1997.
3. Wagner H, Bladt S. Pharmaceutical quality of hypericum extracts. *J of Geriat Psych Neurol* 1994; 7: 65-68.
4. Piovan A, Filippini R, Caniato R, *et al.* Detection of hypericins in the "red glands" of *Hypericum elodes* by ESI-MS/MS. *Phytochemistry* 2004; 65: 411-414.
5. Raffa RB. Screen of receptor and uptake-site activity of hypericin component of St. John's wort reveals sigma receptor binding. *Life Sci* 1998; 62: 265-270.
6. Suzuki O, Katsumata Y, Oya M, *et al.* Inhibition of monoamine oxidase by hypericin. *Planta Med* 1984; 50: 272-274.
7. Demisch L, Holzl J, Gollnik B, *et al.* Identification of selective MAO-type A inhibitors in *Hypericum perforatum* L. (Hyperforat ®). *Pharmacopsychiat* 1989; 22: 194.
8. Lieberman S. Nutraceutical review of St. John's Wort (Hypericum perforatum) for the treatment of depression. *J Woman's Health* 1998; 7: 177-181.
9. Couldwell WT, Gopalkrishna R, Hinton DR, *et al.* Hypericin: a potential antiglioma therapy. *Neurosurgery* 1994; 35: 705-710.
10. Vandenbogaerde A de, Witte P. Antineoplastic properties of photosensitized hypericin (Meeting abstract). *Anticancer Res* 1995; 15: 1757-1758.
11. Alecu M, Ursaciuc C, Halalau F, *et al.* Photodynamic treatment of basal cell carcinoma andsquamous cell carcinoma with hypericin. *Antican Res* 1988; 18: 4651-4654.

12. Takahashi I, Nakanishi S, Kobayashi E, *et al.* Hypericin and pseudohypericin specifically inhibit protein kinase C: possible relation to their antiretroviral activity. *Biochem Biophys Res Commun* 1989; 165: 1207-1212.
13. Mcauliffe V, Gulick R, Hochster H, *et al.* A phase I dose escalation study of synthetic hypericin in HIV infected patients (ACTG 150). *Nat Conf Hum Retroviruses Rel Infect* 1993; Dec 12-16: 159.
14. Vonsover A, Steinbeck KA, Rudich C, *et al.* HIV-1 virus load in the serum of AIDS patients undergoing long term therapy with hypericin. *Int Conf AIDS* 1996; 11: 120 (abstract no. Mo.B.1377).
15. Gulick RM, McAuliffe V, Holden-Wiltse J, *et al.* Phase I studies of hypericin, the active compound in St. John's wort, as an antiretroviral agent in HIV-infected adults. *Ann Int Med* 1999; 130: 510-514.
16. Panossian AG, Gabrielian V, Manvelian K, *et al.* Immunosuppressive effects of hypericin on stimulated human leukocytes: inhibition of the arachidonic acid release, leukotriene B4 and interleukin-1α (production, and activation of nitric oxide formation). *Phytomedicine* 1996; 3: 19-28.
17. Stock S, Holz J. Pharmacokinetic test of (14 C)-labeled hypericin and pseudohypericin from *Hypericum perforatum* and serum kinetics of hypericin in man. *Planta Med* 1991; 57: 61-62.
18. Kerb R, Brockmoller J, Staffeldt B, *et al.* Single-dose and steady-state pharmacokinetics of hypericin and pseudohypericin. *Antimicrob Agents Chemother* 1996; 40: 2087-2093.

Chapter 11

Hyperforin-Acylpholoroglucinol with Significant Antidepressant Potential

11.1 ACYLPHOLOROGLUCINOLS OF *HYPERICUM PERFORATUM* LINN.

According to Erdelmeier, acylphloroglucinols are found in fresh herbs of *H. perforatum*.[1] The main acylphloroglucinol is hyperforin and is highly lipophilic, very photo- and oxygen-labile compound (Fig. 11.1). When exposed to air and light, hyperforin degrades rapidly. Hyperforin is soluble in DMSO and organic solvents.

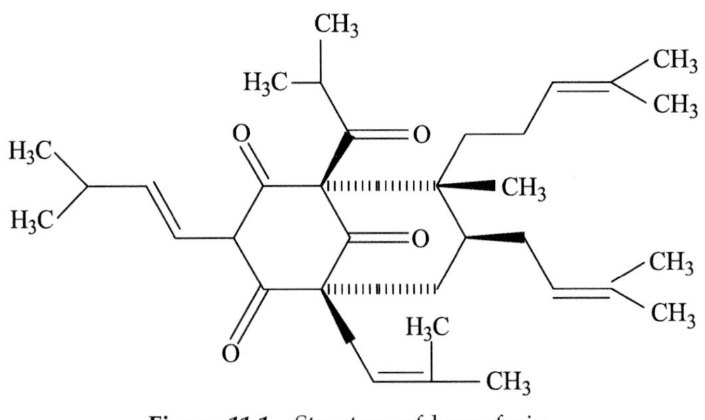

CH_3 H_3C O O CH_3 CH_3 H_3C H_3C CH_3 O CH_3 CH_3 H_3C O CH_3

Figure 11.1 Structure of hyperforin.

Maisenbacher and Kovar, reported adhyperforin (Fig. 11.2) from *H. perforatum*.[2]

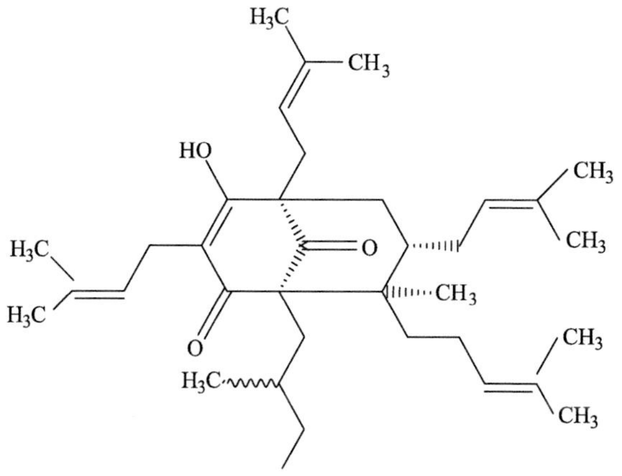

Figure 11.2 Structure of adhyperforin.

According to Wagner and Bladt, wild crafted *H. perforatum* from Germany, dried at 30°C, was found to contain 2.8 percent hyperforin.[3] Rucker et al., reported a new compound containing a hyperforin and a sesquiterpene moiety from leaves and stems of *Hypericum perforatum*. Chatterjee et al., explained that if *H. perforatum* is extracted with supercritical carbon dioxide, higher amount of hyperforin can be obtained.[4]

11.2 BIOSYNTHESIS OF HYPERFORIN

Adam et al., investigated the biosynthesis of hyperforin in *H. perforatum*. The process involves five isoprenoid units, which comes from the deoxyxylulose pathway. The phloroglucinol originates via polyketide pathway.[5] Verotta see acylphloroglucinols have a common 1,3,5-trihydroxy-benzene core of polyketide origin and as lead compounds for pharmacotherapy of degenerative disorders.[6]

11.3 ANTIDEPRESSANT ACTIVITY

11.3.1 Comparative Study of Two *H. perforatum* Extracts (Difference in Hyperforin Content)

Laakmann et al., in a randomized, double-blind, placebo-controlled, multicenter study, investigated the clinical efficacy and safety of two different extracts of *H. perforatum*. The manufacturing process for the two Hypericum preparations was identical, so that they differed only in their hyperforin content. The study included 147 male and female outpatients suffering from mild or moderate depression according to DSM-IV criteria. Following a placebo run-in period of three to seven days, the patients were randomized to one of three treatment groups—During the 42-day treatment period, they received 3 × 1 tablets of either placebo, *H. perforatum* extract

WS 5573 (300 μg, *with a content of 0.5 percent hyperforin*), or *H. perforatum* extract WS 5572 (300 μg, *with a content of 5 percent hyperforin*).

The efficacy regarding depressive symptoms was assessed on days 0, 7, 14, 28, and 42, using the Hamilton Rating Scale for Depression (HAMD, 17-item version) and the Depression Self-Rating Scale (D-S) according to von Zerssen. In addition, the severity of illness was also rated by the investigators on days 0 and 42 using the Clinical Global Impression (CGI) scale. The last observation of patients withdrawn from the trial prematurely was carried forward. At the end of the treatment period (day 42), patients receiving WS 5572 (5 percent hyperforin) exhibited the largest HAMD reduction versus day 0 (10.3 +/− 4.6 points; mean +/− SD), followed by the WS 5573 group (0.5 percent hyperforin; HAMD reduction 8.5 +/− 6.1 points) and the placebo group (7.9 +/− 5.2 points). As regards the change in the HAMD total score between day 0 and treatment end and its relationship to the hyperforin dose, a significant monotonic trend was demonstrated in the Jonckheere-Terpstra test (P = 0.017). In pairwise comparisons, WS 5572 (5 percent hyperforin) was superior to the placebo in alleviating depressive symptoms according to HAMD reduction (Mann-Whitney U-test: P = 0.004), whereas the clinical effects of WS 5573 (0.5 percent hyperforin) and the placebo were descriptively comparable. The results show that the therapeutic effect of *H. perforatum* in mild to moderate depression depends on its hyperforin content.[7]

11.3.2 Inhibition of the Neurotransmitters by Hyperforin

A. Wonnermann et al., explained a new mechanism for antidepressant action of *H. perforatum* while investigating the possible role of hyperforin in antidepressant activity. The researchers using a standardized extract of *H. perforatum* (LI 160) demonstrated that the extract not only inhibits the uptake of serotonin, or epinephrine, dopamine, but also inhibits gamma amino butyric acid and L-glutamate. The uptake inhibition of hyperforin was compared with that of Na(+)-ionosphere monoensin and ouabain. In the presence of amiloride and MIA (non-selective inhibitors of Na(+)/H(+) antiporters and Na(+)-channels), benzamil (Na(+) channel blocker) and EIPA (inhibitor of Na(+)/H(+) exchangers), they proved reduced gamma amino butyric acid and L-glutamate by hyperforin. On the other hand, monoesin or ouabain inhibition were not affected. Thus they concluded that hyperforin inhibits serotonin uptake by elevating free intracellular Na(+).[8]

B. Muller et al., suggested that hyperforin not only inhibits the neuronal uptake of serotonin, norepinephrine and dopamine like many other antidepressants, but also inhibits gamma-amino butyric acid and L-glutamate uptake. This broad-spectrum effect is obtained by an elevation of the intracellular Na(+) concentration. It is probably due to activation of sodium conductive pathways (not yet established but most likely ionic channels). The study claimed hyperforin as the first member of a new class of compounds with a preclinical antidepressant profile having a novel and distinct mechanism of action.[9]

11.3.3 Physicochemical Interaction of Hyperforin with Specific Membrane Structures

Eckert and Muller claimed that hyperforin affects several ionic conductance mechanisms in brain cells although the mechanism of action is still is awaited. The study showed the effects of hyperforin on the fluidity of crude brain membranes from young guinea pigs. Fluidity measurements were performed with three different fluorescent probes. Diphenylhexatriene and trimethylammonium-diphenylhexatriene anisotropy measurements were inversely correlated with the flexibility of fatty acids in the membrane hydrocarbon core and in the hydrophilic area of membrane phospholipids, respectively. The ratio of pyrene excimer to monomer fluorescence intensities was used as an indicator of membrane annular and bulk fluidity.

Incubation of brain membranes with relatively high concentrations of hyperforin sodium salt (10 μM/l) resulted in increased diphenylhexatriene and decreased trimethylammonium-diphenylhexatriene anisotropy, respectively, indicating that hyperforin modifies specific membrane structures in different ways. It decreases the flexibility of fatty acids in the membrane hydrocarbon core, but fluidizes the hydrophilic region of membrane phospholipids. Interestingly, relatively low concentrations of hyperforin (0.3 μM/l) significantly decreased the annular fluidity of lipids close to membrane proteins.[10]

11.3.4 Similarity in Action of Fluoxetine and Hyperforin

Hoye et al., determined if fluoxetine had similar effects on glutamate reuptake inhibition as hyperforin. Effects of the known glutamate reuptake inhibitor, aminocaproic acid were examined in order to establish a framework through which to compare the effects of selective serotonin reuptake inhibitors on glutamate reuptake inhibition. Fluoxetine exhibited the largest increase in duration, hyperforin showed the smallest increase, and the combination of fluoxetine and hyperforin exhibited an increase in EPSP duration that was between that of the two Selective Serotonin Reuptake Inhibitors (SSRIs) alone. It was concluded that fluoxetine had a greater effect on the reuptake inhibition of glutamate than hyperforin. An unexpected result came when fluoxetine and hyperforin were used together in that the averaging affect of the two chemicals presents that possibility that combining these two drugs is safer than using fluoxetine alone, in regards to glutamate reuptake inhibition.[11]

11.4 COGNITIVE (ANTI-AMNESIC) AND NEUROPROTECTIVE ACTIVITIES

11.4.1 Cognitive

Klusa et al., evaluated effect of *H. perforatum* extract and sodium salt of hyperforin in rat avoidance test. In a conditioned avoidance response test on rats, oral daily administration of hyperforin (1.25 mg/kg/day) or of the extract of *H. perforatum* (50 mg/kg/day) before the training sessions significantly improved learning ability from the second day onward until day seven. The memory attained during seven days of treatment was retained for nine days without further treatment. In a passive avoidance response test, a single oral dose of hyperforin not only resulted in improved memory acquisition but completely reversed scopolamine induced amnesia. Single dose of *H. perforatum* extract (50 mg/kg/day) did not produce significant effect in passive avoidance test in the animals.[12]

11.4.2 Neuroprotective

In the work of Dinamarca et al., the effect of hyperforin on Abeta-induced spatial memory impairments and on Abeta neurotoxicity was determined. Hyperforin decreased amyloid deposit formation in rats injected with amyloid fibrils in the hippocampus; decreased the neuropathological changes and behavioral impairments in a rat model of amyloidosis; and prevented Abeta-induced neurotoxicity in hippocampal neurons both from amyloid fibrils and Abeta oligomers, avoiding the increase in reactive oxidative species associated with amyloid toxicity. The evidence suggests that hyperforin may be useful to decrease amyloid burden and toxicity in Alzheimer's patients, and may be a putative therapeutic agent to fight the disease.[13]

Hyperforin has been shown to have neuroprotective effects against Alzheimer's disease neuropathology, including the ability to disassemble amyloid-beta aggregates *in vitro*, decrease astrogliosis and microglia activation, as well as improve spatial memory *in vivo*.[14]

11.4.3 N-methyl-D-aspartate Antagonistic

Kumar et al., reported that hyperforin also has N-methyl-D-aspartate-antagonistic effects. Hyperforin (10 μM) was found to inhibit the N-methyl-D-aspartate-induced calcium influx into cortical neurons. In rat hippocampal slices, hyperforin inhibited the N-methyl-D-aspartate-receptor-mediated release of choline from phospholipids. Hyperforin also antagonized the increase of water content in freshly isolated hippocampal slices, and it counteracted, at 3 and 10 μM, the increase of water content induced by N-methyl-D-aspartate. Hyperforin was inactive, however, in two *in vivo* models of brain edema formation, middle cerebral artery occlusion and water intoxication in mice.[15]

11.4.4 Antidementia

Cepra et al., reported effect of hyperforin derivative (IDN5706, tetrahydrohyperforin), a semi-synthetic derivative of the *H. perforatum*, on the brain neuropathology, learning and memory in a double transgenic (APPswe, PS-1dE9) mouse model of Alzheimer's disease. Results indicate that, IDN5706 alleviates memory decline induced by amyloid-beta (Abeta) deposits as indicated by the Morris water maze paradigm. Moreover, the analysis of Abeta deposits by immunodetection and thioflavin-S staining of brain sections, only reveals a decrease in the frequency of the larger-size Abeta deposits, suggesting that IDN5706 affected the turnover of amyloid plaques. Immunohistochemical analysis, using GFAP and n-Tyrosine indicated that the hyperforin derivative prevents the inflammatory astrocytic reaction and the oxidative damage triggered by high Abeta deposit levels.[16]

11.5 ANXIOLYTIC ACTIVITY

Zanoli et al., investigated pharmacological activity of hyperforin acetate. In the forced swimming test, triple administration of hyperforin resulted in significant reduction in immobility time of rats. In the light-dark test, hyperforin acetate demonstrated anxiolytic effect but smaller as compared to diazepam. Hyperforin acetate demonstrated muscle relaxant activity as evident from reduced locomotion in rats. Acute as well as repeated doses of hyperforin acetate did not altered pentobarbital sleeping time in rats.[17]

11.6 ANTITUMOUR ACTIVITY

Schempp et al., reported inhibition of tumor cell growth by hyperforin. The effect of hyperforin was analyzed in a range of human and rat tumor cell lines *in vivo* and *in vitro*. Hyperforin inhibited human and rat tumor cell growth (IC50 3-15 μM). Hyperforin resulted in a dose-dependent generation of apoptotic oligonucleosomes, typical DNA–laddering and structural changes characteristic of apoptosis. In a breast cancer cell line (MT-450), hyperforin increased the activities of enzymes caspase 9 and caspase 3.

In MT-450 cells, hyperforin caused damage to mitochondria. Furthermore hyperforin caused the release of cytochrome C from isolated mitochondria. When hyperforin was compared with paclitaxel *in vivo,* both inhibited growth of MT-450 cells in rats. However hyperforin showed no signs of acute toxicity.[18]

11.7 ANTI-INFLAMMATORY ACTIVITY

Isabella et al., tested whether hyperforin while inhibiting leukocyte elastase activity, might also be effective in containing both polymorphonuclear neutrophil leukocyte recruitment and unfavorable eventual tissue responses.

The results show that, without affecting *in vitro* human polymorphonuclear neutrophil viability and chemokine-receptor expression, hyperforin (as stable dicyclohexylammonium salt) was able to inhibit in a dose-dependent manner their chemotaxis and chemoinvasion (IC50 = 1 µM for both); this effect was associated with a reduced expression of the adhesion molecule CD11b by formyl-Met-Leu-Phe-stimulated neutrophils and block of leukocyte elastase-triggered activation of the gelatinase matrix metalloproteinase-9. Polymorphonuclear neutrophil-triggered angiogenesis is also blocked by both local injection and daily IP administration of the hyperforin salt in an interleukin-8-induced murine model.[19]

11.8 SCOPE OF HYPERFORIN IN DERMATOLOGICAL RESEARCH

Hyperforin has recently been identified as a specific transient receptor potential (TRPC6) activator, Müller investigated the contribution of TRPC6 to keratinocyte differentiation and proliferation. Like the endogenous differentiation stimulus high extracellular Ca(2+) concentration ([Ca(2+)](o)), hyperforin triggers differentiation in HaCaT cells and in primary cultures of human keratinocytes by inducing Ca(2+) influx via TRPC6 channels and additional inhibition of proliferation. Knocking down TRPC6 channels prevents the induction of Ca(2+)- and hyperforin-induced differentiation. More important, TRPC6 activation is sufficient to induce keratinocyte differentiation similar to the physiological stimulus [Ca(2+)](o). Therefore, TRPC6 activation by hyperforin may represent a new innovative therapeutic strategy in skin disorders characterized by altered keratinocyte differentiation.[20]

11.9 ALCHOLISM AND HYPERFORIN

Perfumi et al., compared the effect on ethanol intake in alcohol-preferring rats of two *Hypericum perforatum* extracts—a methanolic extract containing 0.3 percent hypericin and 3.8 percent hyperforin (HPE1) and a CO_2 extract (HPE2) with 24.33 percent hyperforin and very low hypericin content. Freely feeding and drinking rats were offered 10 percent ethanol 2 hr/day and HPE were given intragastrically 1 hr before access to ethanol. Both extracts dose-dependently reduced ethanol intake, HPE2 being about eight times more potent than HPE1. Food and water intakes were not affected by doses that reduced ethanol intake. HPE2, unlike HPE1, reduced blood-alcohol levels (BAL) at doses of $\geq$ 31.2 mg/kg, whereas the dose of 15.6 mg/kg, which reduced ethanol intake, did not significantly modify BAL; blood-acetaldehyde levels were not increased.

Intraperitoneal pretreatment with the sigma-1 receptor antagonist NE-100 (0.25 mg/kg) did not affect inhibition of ethanol intake induced by HPE1 (250 mg/kg) or HPE2 (125 mg/kg), but stopped the effect of both extracts in the FST. The results indicate that HPE2 inhibits ethanol intake more potently than HPE1; the higher potency of HPE2 parallels the hyperforin content, suggesting that hyperforin may have an important role in reducing ethanol intake.[21]

11.10 PHARMACOKINETICS

Biber et al., investigated the pharmacokinetics of hyperforin in healthy volunteers. After administration of a 300 mg tablet of extract of *H. perforatum* containing 14.8 mg hyperforin, a maximum plasma level of approximately 150 µg/ml (280 µM) hyperforin was reached after 3.5 hr. The oral bioavailability of hyperforin in doses up to 30 mg (i.e. 600 mg St. John's wort extract) was high. The half-life of hyperforin was 9 hr and the mean residence time 12 hr. No accumulation of hyperforin occurred with repeated dosing. Estimated steady state plasma concentrations with 3 × 300 mg extract per day were approximately 100 µg/ml or 180 µM.[22]

Cui et al., investigated *in vitro* metabolism profile of hyperforin using liver microsomes from male and female Sprague-Dawley rats, with or without induction by phenobarbital or dexamethasone. Four major Phase I metabolites, called 19-hydroxyhyperforin, 24-hydroxyhyperforin, 29-hydroxyhyperforin and 34-hydroxyhyperforin, were isolated by high performance liquid chromatography and identified by mass spectrometry and NMR. Results suggest that hydroxylation is a major biotransformation of the hyperforin pathway in rat liver and that inducible cytochrome P450 3A (CYP450 3A) and/or CYP2B may be the major cytochrome P450 isoforms catalyzing these hydroxylation reactions.[23]

11.11 DRUG INTERACTIONS

Cantoni et al., reported that hyperforin contributes to the hepatic CYP3A-inducing effect of *H. perforatum* extract in the mouse. A hydroalcoholic extract containing 4.5 percent hyperforin was given at a dose of 300 mg/kg b.i.d. for 4 and 12 days. Hyperforin, its main phloroglucinol component, was given as dicyclohexylammonium salt (18.1 mg/kg, b.i.d.) on the basis of its content in the extract, to ensure comparable exposure to hyperforin. The extract increased hepatic erythromycin-N-demethylase activity, which is cytochrome P450 enzyme (CYP) 3A-dependent, about 2.2-fold after four days of dosing, with only slightly greater effect after 12 days (2.8 times controls). Hyperforin also increased erythromycin-N-demethylase activity within four days, almost the same extent as the extract (1.8 times the activity of controls), suggesting that it behaves qualitatively and quantitatively like the extract as regards induction of CYP3A activity. This effect was confirmed by Western blot analysis of hepatic CYP3A expression. Exposure to hyperforin at the end of the four days' treatment was still similar to that with extract of *H. perforatum*, although it was variable and lower than after the first dose in both cases, further suggesting that hyperforin plays a key role in CYP3A induction by the extract of *H. perforatum* in the mouse. Standardization of the extracts based on the hyperforin content can be proposed for further evaluation of their potential action on first-pass metabolism and clearance of co-administered CYP3A substrates.[24]

Watkins et al., in a study showed that hyperforin induces the expression of numerous drug metabolisms and excretion genes in primary human hepatocytes. Researchers determined the crystal structure of hyperforin in complex with the ligand binding domain of human PXR. Hyperforin induces conformational changes in PXR's ligand binding pocket relative to structures of human PXR elucidated previously and increases the size of the pocket by 250 A(3). It was found that the mutation of individual aromatic residues within the ligand binding cavity changes PXR's response to particular ligands. Together, these results demonstrate that PXR uses structural flexibility to expand the chemical space it samples and that the mutation of specific residues within the ligand binding pocket of PXR tunes the receptor's response to ligands.[25]

REFERENCES

1. Erdelmeier CAJ. Hyperforin, possibly the major non-nitrogenous secondary metabolite of *Hypericum perforatum* L. *Pharmacopsychiatry* 1998; 31: 2-6.
2. Maisenbacher P, Kovar KA. Adhyperforin - a homologue of hyperforin from *Hypericum perforatum*. *Planta Med* 1992; 58: 291-293.
3. Wagner H, Bladt S. Pharmaceutical quality of hypericum extracts. *J of Geriat Psych Neurol* 1994; 7: 65-68.
4. Chatterjee SS, Noldner M, Koch E, *et al.* Antidepressant activity of *Hypericum perforatum* and hyperforin-the neglected possibility. *Pharmacopsychiatry* 1998; 31: 7-15.
5. Adam P, Arigoni D, Bacher A, *et al.* Biosynthesis of hyperforin in *Hypericum perforatum*. *Med Chem* 2002; 45: 4786-4793.
6. Verotta L. Are phlorogucinols as lead structures for the treatment of degenerative diseases? Dipartimento di Chimica Organica e Industriale, Universita degli Studi di Milano, via Venezian 21, 20133 Milano, Italy.
7. Laakmann G, Schule C, Baghai T, *et al.* St. John's wort in mild to moderate depression: the relevance of hyperforin for the clinical efficacy. *Pharmacopsychiatry* 1998; 31: 54-59.
8. Singer A, Wonnemann M, Muller WE. Hyperforin, a major antidepressant constituent of St. John's wort, inhibits serotonin uptake by elevating free intracellular $Na^{+}1$. *Pharmacol Exp Ther* 1999; 290: 1363-1368.
9. Muller WE, Singer A, Wonnemann M. Hyperforin-antidepressant activity by a novel mechanism of action. *Pharmacopsychiatry* 2001; 34: 98-102.

10. Eckert GP, Muller WE. Effects of hyperforin on the fluidity of brain membranes. *Pharmacopsychiatry* 2001; 34: 22-25.
11. Hoye A, Mitchell D, Tucker A. Fluoxetine and hyperforin appear to act like a known glutamate reuptake inhibitor by increasing EPSP duration in the crayfish neuromuscular junction. *Pioneer Neurosci* 2002; 3: 41-44.
12. Klusa V, Germane S, Noldner M, *et al.* Hypericum extract and hyperforin: memory-enhancing properties in rodents. *Pharmacopsychiatry* 2001; 34: 61-69.
13. Dinamarca MC, Cerpa W, Garrido J, *et al.* Hyperforin prevents beta-amyloid neurotoxicity and spatial memory impairments by disaggregation of Alzheimer's amyloid-beta-deposits. *CNS Drug Rev* 2004; 10: 203-218.
14. Griffith TN, Varela-Nallar L, Dinamarca MC, *et al.* Neurobiological effects of Hyperforin and its potential in Alzheimer's disease therapy. *Curr Med Chem* 2010; 17: 391-406.
15. Kumar V, Mdzinarishvili A, Kiewert C, *et al.* NMDA receptor-antagonistic properties of hyperforin, a constituent of St. John's Wort. *J Pharmacol Sci* 2006; 102: 47-54.
16. Cerpa W, Hancke JL, Morazzoni P, *et al.* The hyperforin derivative IDN5706 occludes spatial memory impairments and neuropathological changes in a double transgenic Alzheimer's mouse model. *Curr Alzheimer Res* 2010; 7: 126-133.
17. Zanoli P, Rivasi M, Baraldi C, *et al.* Pharmacological activity of hyperforin acetate in rats. *Behav Pharmacol* 2002; 13: 654-651.
18. Schempp CM, Kirkin V, Simon-Haarhaus B, *et al.* Inhibition of tumor cell growth by hyperforin, a novel anticancer drug from St. John's wort that acts by induction of apoptosis. *Onceogene* 2002; 21: 1242-1250.
19. Dell'Aica I, Niero R, Piazza F, *et al.* Hyperforin Blocks Neutrophil Activation of Matrix Metalloproteinase-9, Motility and Recruitment, and Restrains Inflammation-Triggered Angiogenesis and Lung Fibrosis. *J Pharmacol Exp Ther* 2007; 321: 492-500.
20. Müller M, Essin K, Hill K, *et al.* Specific TRPC6 channel activation, a novel approach to stimulate keratinocyte differentiation. Specific TRPC6 channel activation, a novel approach to stimulate keratinocyte differentiation. *J Biol Chem* 2008; 283: 33942-33954.
21. Perfumi M, Panocka I, Ciccocioppp R, *et al.* Effects of a methanolic extract and a hyperforin-enriched CO_2 extract of *Hypericum perforatum* on alcohol intake in rats. *Alcohol* 2001; 36: 199-206.
22. Biber A, Fischer H, Romer A, *et al.* Oral bioavailability of hyperforin from hypericum extracts in rats and human volunteers. *Pharmacopsychiatry* 1998; 31: 36-43.
23. Cui Y, Ang CY, Beger RD, *et al.* *In vitro* metabolism of hyperforin in rat liver microsomal systems. *Drug Metab Dispos* 2004; 32: 28-34.

24. Cantoni L, Rozio M, Mangolini A, *et al.* Hyperforin Contributes to the Hepatic CYP3A-Inducing Effect of *Hypericum perforatum* Extract in the Mouse. *Biotransform Toxicokin.* Society of Toxicology, 2003.
25. Watkins RE, Maglich JM, Moore LB, *et al.* A crystal structure of human PXR in complex with the St. John's wort compound hyperforin. *Biochemistry* 2003; 42: 1430-1438.

Chapter 12

Survey of Indian Medicinal Plants for Alkaloids and Pharmacology

12.1 INTRODUCTION

Herbal drugs constitute a major share of all the officially recognized systems of health in India viz. Ayurveda, Yoga, Unani, Siddha, Homeopathy and Naturopathy, except for Allopathy.[1] India is one of the 12 mega diversity countries in the world so it has a vital stake in conservation and sustainable utilization of its biodiversity resources. Plant secondary metabolites have been of interest to man for a long time due to their pharmacological relevance.[2]

Indian medicinal plants have been studied for pharmacological activity in recent years. To understand the mechanism of action, researchers worked at molecular levels and several significant phytochemicals have been isolated. The present chapter compiles data on promising phytochemicals from Indian medicinal plants that have been tested in various disease models using modern scientific methodologies and tools.[3]

India is endowed with a rich variety of medicinal plants. Several useful drugs have been isolated from Indian medicinal plants, predominantly alkaloids. Sterneur isolated morphine from *Papaver somniferum* Linn. (opium poppy) and showed the medical profession that certain phytochemicals produced in plant cells are responsible for pharmacological activity.

Later other alkaloids isolated from opium poppy were investigated for their pharmacological activities. Codeine showed antitussive activity and papaverine antispasmodic activity. Opium based extracts have been used for various pharmacological activities, and a number of alkaloids distributed in the plant have different pharmacological activities.[4]

The chapter is dedicated to studies done on alkaloids reported from Indian medicinal plants. With revised interest in traditional medicine, scientists are looking for cost effective and potent remedies from medicinal plants. Scientific information on medicinally useful chemical constituents of medicinal plants in scattered and there is urgent need to assemble it at one point.

12.2 MATERIALS AND METHODS

The keywords for the present chapter are alkaloid, stereochemistry, pharmacology and Indian Medicinal Plants (IMP). The search was done using Google, Pub Med, Ayush Health portal, Annotated Bibliography of Indian Medicine (ABIM) and data bank on Indian Medicinal Plants maintained by department of Central Council for Research in Ayurveda and Siddha (CCRAS) updated until January, 2013. Standard periodicals and books (including ancient Materia Medica) on Indian Medicinal Plants were consulted for cross-reference of data generated from search engines mentioned above. Alkaloids were classified according to their action human systems.

12.3 RESULTS

12.3.1 Antimicrobial

Imidazole alkaloid, chaksine (Fig. 12.1), isolated from *Cassia absus* Linn. (Fabaceae) has antibacterial activity.[5] In a test for antifungal activity, chaksine iodide at 0.5 percent inhibited all fungi tested.[6] Isochaksine has similar activities to chaksine but generally at higher doses.[7] Ditamine or ditanin an alkaloid present in *Alstonia scholaris* (L.) R.Br. (Apocynaceae) is reported to be antimalarial. Ditamine possesses antiperiodic properties equal to the best quinine sulfate.[8]

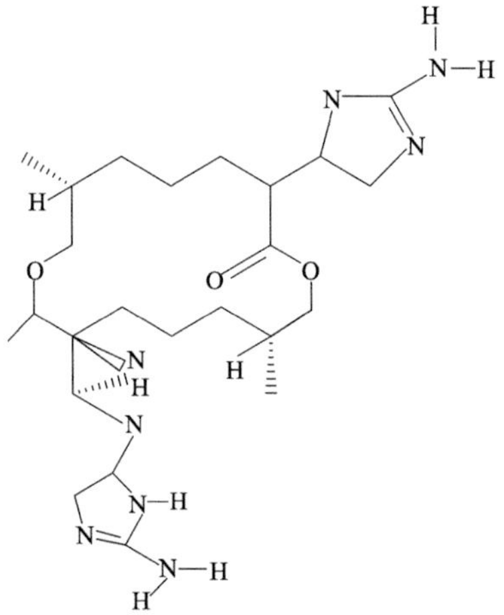

Figure 12.1 Structure of chaksine.

A new carbazole alkaloid, clausenol (Fig. 12.2), isolated from an alcoholic extract of the stem bark of *Clausena anisata* (Willd) Hook (Rutaceae) was found to be active against Gram-positive and Gram-negative bacteria and fungi.[9]

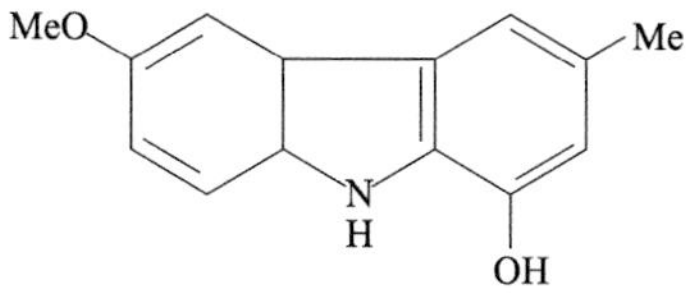

Figure 12.2 Structure of clausenol.

12.3.2 Digestive System

12.3.2.1 Antidiarrheal

Antidiarrheal activity of piperine, the principle alkaloid of *Piper longum* and *Piper nigrum* L. (Piperaceae) was investigated against diarrhea induced by castor oil, magnesium sulfate and arachidonic acid in rats. Piperine (Fig. 4.3) significantly inhibited diarrhea produced by the above laxatives and aperients at a dose of 8 and 32 mg/kg PO dose. Piperine inhibited castor oil induced enter pooling explaining prostaglandin inhibiting effect.[10]

12.3.2.2 Hepatoprotective

Leaf extract *of Ricinus communis* L. (Euphorbiaceae) was evaluated for hepatoprotective, choleretic and anticholestatic activity. In a preliminary test with albino rats, an ethanol extract showed significant protection against galactosamine-induced hepatic damage. It also showed dose-dependent choleretic and anticholestatic activity, and hepatoprotective activity as judged by hepatocytes isolated from paracetamol-treated rats. On fractionation of the ethanol extract, maximum activity was localized in the butanol fraction. Subsequent chromatographic fractionation and testing in the galactosamine model led to the isolation of two active fractions, which in turn yielded two pure compounds: ricinine (Fig. 12.3) and N-demethyl-ricinine.[11]

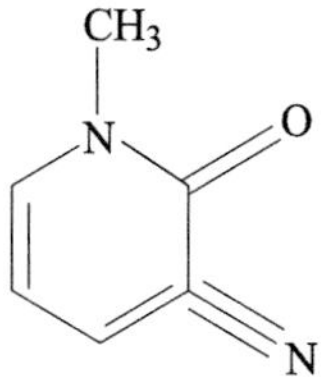

Figure 12.3 Structure of ricinine.

12.3.3 Respiratory System

12.3.3.1 Bronchodilator

Saussurine, an alkaloid isolated from *Saussurea lappa* (Decne.) C.B. Clarke (Asteraceae) is a potent bronchodilator.[12] It has depressant action on the involuntary muscle fibers of the bronchioles and gastrointestinal tract. Saussurine tartrate has been shown to produce definite relaxation of the bronchioles in the same way as adrenaline.[13] Aegeline, an alkaloid of *Aegle marmelos* has been reported to be useful in bronchial asthma.[14]

Moringine (Fig. 12.4) from *Moringa olifera* Lam. (Moringaceae) is reported to be a cardiac stimulant. It has a bronchodilator, depressant action on smooth muscle fibers and an inhibitory effect on the tone and movements of the intestine in rabbits and guinea pigs. Experiments conducted with tylophorine (Fig. 12.5), a phenanthroindalizidine alkaloid, present in *Tylophora asthmatica* (L.f.) Wight & Arn. (Asclepiadaceae) in various animal models have shown significant anti-inflammatory, anti-anaphylactic and anti-spasmodic activities. Bronchodilator activity *of Tylophora asthmatica* is attributed to tylophorine.[15]

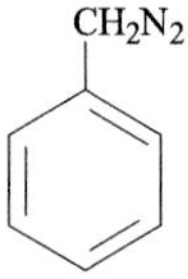

Figure 12.4 Structure of moringine.

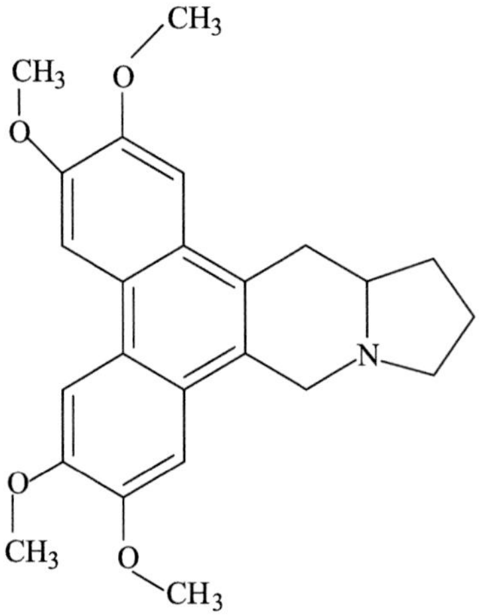

Figure 12.5 Structure of tylophorine.

12.3.3.2 Mast cell stabilizer

Lycoriside, at lower concentrations (1-20 μg/ml), *in vitro*, produced statistically significant protection against Tween 80-induced degranulation, as also to sensitized mast cells challenged with an antigen (horse serum). It also provided protection against the compound 48/80-induced degranulation of mast cells when administered *in vivo* in a dose of 1-5 mg/kg, PO.[16]

12.3.4 Cardiovascular System

12.3.4.1 Cardiac tonic

Aegeline is reported to have a cardiac tonic activity. Saussurine has a positive inotropic action on the heart muscle.[17] The hydrochloride of alkaloid, evolvine present in *Evolvulus alsinoides* Linn. (Convolvulaceae) was reported to exhibit lobeline-like action on the cardiovascular system. In cats, the drug demonstrated sympathomimetic activity. Blood pressure remained elevated for a longer duration as compared to adrenaline. Increase in peripheral pressure was observed on local injection of the drug.[18] Moringine is reported to be cardiac stimulant.[4]

12.3.4.2 Hemostatic

Achilleine, an alkaloid *Achillea millefolium* L. (Asteraceae) is considered to be active hemostatic. It reduces blood-clot in rabbits.[19]

12.3.5 Central Nervous System

12.3.5.1 Analgesic

Cerpegin (Fig. 12.6), a novel furopyridine alkaloid isolated from the plant *Cerpegia juncea* Roxb. (Asclepiadaceae) was subjected to various pharmacological investigations. A dose-related analgesic effect was observed in mice. Cerpegin did not produce any autonomic or behavioral changes up to a dose of 200 mg/kg but doses of more than 400 mg/kg produced excitation and later convulsions in mice.[20] Total alkaloids of *E. alba.* have been reported to have significant analgesic activity in all the different models of analgesia used.[21]

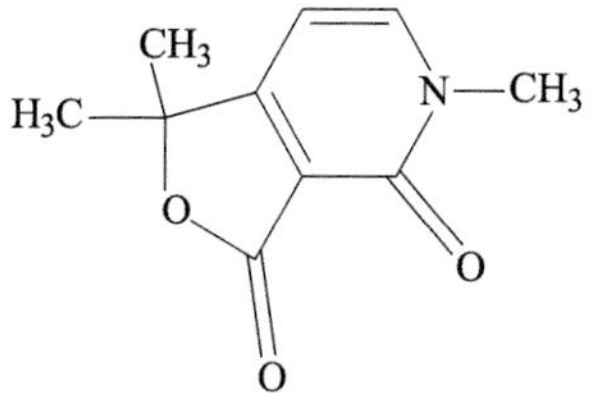

Figure 12.6 Structure of cerpegin.

12.3.5.2 Anti-inflammatory

Premnazole (Fig. 12.7), an isoxazole alkaloid isolated from *Premna integrifolia* L. and *Gmelina arborea* L. (Verbenaceae) demonstrated significant anti-inflammatory activity in reducing cotton pellet-induced granuloma formation in rats. The anti-inflammatory activity was comparable to that of phenylbutazone.[22]

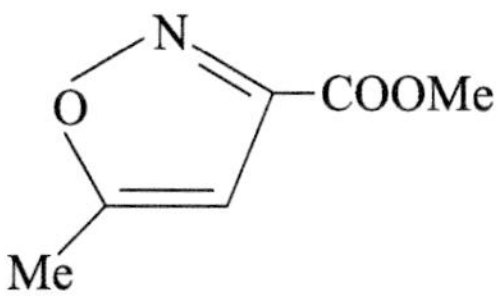

Figure 12.7 Structure of premnazole.

The water-soluble alkaloid, achyranthine isolated from *Achyranthes aspera* L. (Amaranthaceae) was screened for its anti-inflammatory and anti-arthritic activity against carrageenin-induced foot oedema, granuloma pouch, formalin induced arthritis and adjuvant arthritis in rats. It showed significant anti-inflammatory activity in all the four models used but was less active than phenylbutazone and betamethasone. Incidence of gastric ulcers was maximum with betamethasone and minimum with achyranthine.[23]

Anti-inflammatory activity of crotalaburnine or anacrotine (Fig. 12.8) isolated from *Crotalbria laburnifolia* Linn. (Fabaceae) was investigated against several models of inflammation. The effect was compared with the activity of hydrocortisone, phenylbutazone, sodium salicylate and cyproheptadine against different types of inflammation. Crotalaburnine (40 mg/kg SC × 5 alternate days) had no significant inhibitory effect against formalin-induced arthritis, while hydrocortisone (40 mg/kg SC × 10 days) was effective from the fifth day onwards. Against carrageenin-induced oedema both crotalaburnine (10 mg/kg SC) and phenylbutazone (100 mg/kg oral) produced a similar degree of inhibition. Hydrocortisone (10 mg/kg SC) produced slightly greater inhibition.

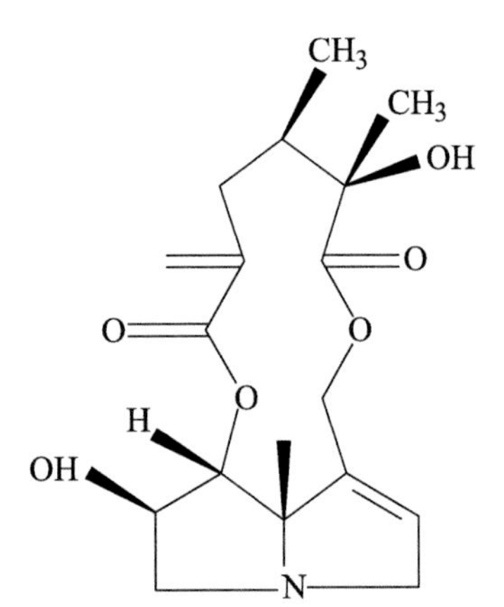

Figure 12.8 Structure of crotalaburnine (anacrotine).

In normal rats crotalaburnine (10 mg/kg SC), phenylbutazone (100 mg/kg oral) and sodium salicylate (500 mg/kg IP) inhibited hyaluronidase-induced oedema. However, in adrenalectomized rats, there was a reduction of the inhibitory effect of sodium salicylate but not of phenylbutazone or crotalaburnine. Crotalaburnine (40 mg/kg SC and 30 mg/kg IP, respectively) was ineffective against 5-hydroxytryptamine- and dextran-induced oedema but against bradykinin- and prostaglandin-induced oedema (in a dose of 20 mg/kg IP) it was quite effective.

In a parallel series cyproheptadine (10 mg/kg oral and IP, respectively) produced significant inhibition of 5-hydroxytryptamine- and dextran-induced oedema, while phenylbutazone (100 mg/kg IP) failed to produce any significant inhibition of prostaglandin-induced oedema. Against cotton-pellet granuloma crotalaburnine, in half the dose of hydrocortisone, produced a similar inhibition while phenylbutazone produced much greater inhibition in five times the dose of crotalaburnine given orally.[24]

12.3.5.3 Central nervous system depressant

Chaksine suppressed the respiratory, vasomotor and thermostat mechanism of the brain.[6]

12.3.6 Reproductive System

12.3.6.1 Antifertility

Solasodine (Fig. 12.9), an alkaloid of *Solanum xanthocarpum* L. (Solanaceae) demonstrated antifertility activity in males and dogs. The alkaloid on oral administration resulted in inhibition of spermatogenesis and sperm motility. The mechanism of action of solasodine may be attributed to the inhibition of testosterone release.[25]

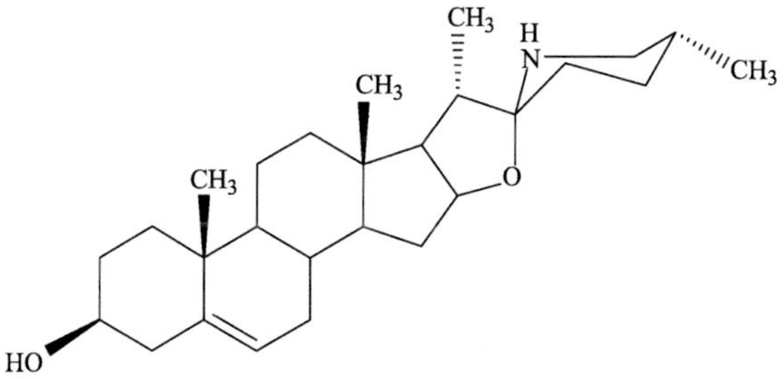

Figure 12.9 Structure of solasodine.

12.3.6.2 Androgenic

In a study, the total alkaloidal content (mucuadine, mucuadinine, mucucuadinine, pruriendine, mucunine, mucunadine and nicotine) of *Mucuna prurita* L. (Fabaceae) demonstrated increased spermatogenesis in albino rats.[26]

12.3.7 Anticancer

Ungeremine, criasbetaine, crinafoline and crinafolidine are reported to have antitumor activity.[27,28] Ungeremine is reported to contribute to growth-inhibiting and cytotoxic effects of lycorine.[27] *In vitro* inhibition studies on the human leukemic leukocyte TS with the phenanthroindolizidine alkaloids, pergularinine and tylophorinidine (Fig. 12.10) isolated from *Pergularia pallida* Wight. & Arn. (Apocynaceae) were conducted for the preliminary screening tests for their antitumor activity. The leukemic leukocyte enzyme activity was potently inhibited by pergularinine (PGL) and tylophorinidine (IC50 = 50 μM) in both types of leukemias.[29]

Pergularinine, tylophorinidine and deoxytubulosine isolated from the Indian medicinal plants *P. pallida* and *Alangium lamarckii* Thw. (Alangiaceae) respectively, inhibited (IC50 = 50 μM) the elevated TS activity of leukocytes in cancer patients with clinically diagnosed chronic myelocytic leukemia (n = 10), acute lymphocytic leukemia and metastatic solid tumors.[30]

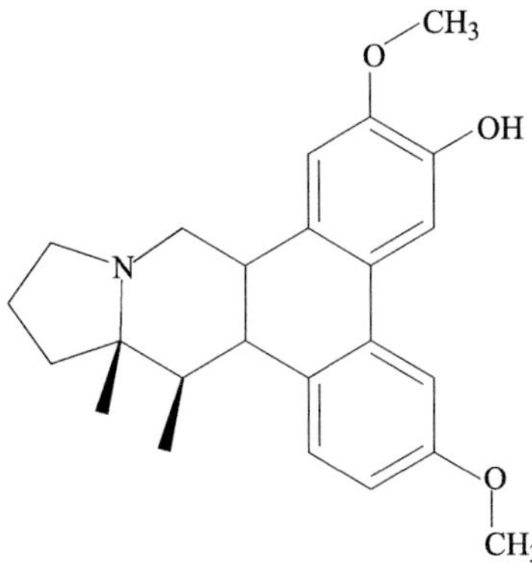

Figure 12.10 Structure of tylophorinidine.

In vitro inhibition studies on the human leukemic leukocyte TS with the phenanthroindolizidine alkaloids pergularinine and tylophorinidine, isolated from *P. pallida* were conducted for the preliminary screening tests for their antitumor activity. The leukemic leukocyte enzyme activity was potently inhibited by pergularinine and tylophorinidine (IC50 = 50 μM) in both types of leukemias.[30]

Ethanolic extracts of six Indian medicinal plants, piperine, guggulsterone E and guggulsterone Z were tested for cytotoxicity using brine shrimp lethality test. *Piper longum* showed most potent cytotoxic activity. Piperine, guggulsterone E and guggulsterone Z showed potent activity with LC50 2.4, 8.9 and 4.9, respectively.[31]

The total alkaloid fraction of the methanolic extract of *Solanum pseudocapsicum* Linn. (Solanaceae) leaves were tested for its *in vivo* antitumor activity against Dalton's Lymphoma Ascites model in mice. The total alkaloid fraction at 2.5 and 5.0 mg/kg body weight doses exhibited antitumor activity as revealed by the significant increase in the mean survival time and the percentage increase in the life span of tumor bearing mice.[32] O-methylsolanocapsine (Fig. 12.11) isolated from *S. pseudocapsicum* leaves possess strong cytotoxic properties.[33]

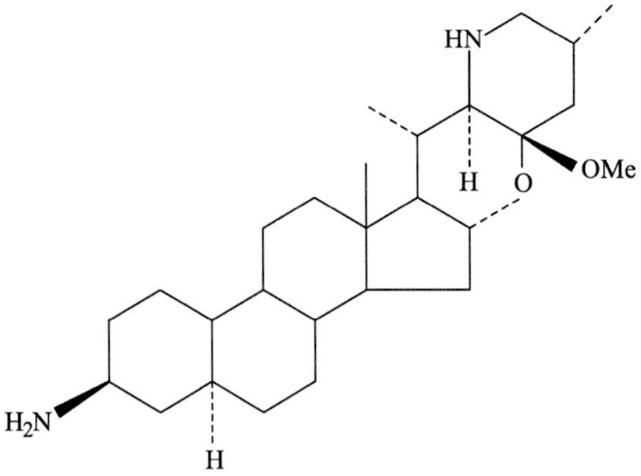

Figure 12.11 Structure of O-methylsolanocapsine.

12.4 DISCUSSIONS

Research has revealed that information related to alkaloids reported from Indian medicinal plants was lacking in various aspects. The major lacuna was in stereochemistry and pharmacological studies. Some alkaloids like aconitine, atropine, berberine, codeine, colchicine, conessine, emetine, morphine, nicotine, strychnine, piperine, quinine, reserpine, are well-defined drugs and data is widely available regarding pharmacological studies.

Data search on alkaloids yielded useful information regarding pharmacological activities. Texts obtained from ancient Materia Medica like the "Pharmacographia Indica" explains isolation procedures of active principles of crude Indian Medicinal Plants. "Materia Medica of Indian Medicinal Plants and Therapeutics" gives a good account of monographs on crude Indian drugs. Work at scattered places describes isolation procedures and pharmacological action of the active principle. For example, saussurine, alkaloid of *Sausuurea lappa* has been described as a bronchodilator, but details regarding animal or human testing are lacking. The preliminary data is useful for carrying out applied research.

"Indian Materia Medica" and "Drugs of India" were significant for obtaining leads from Indian Medicinal Plants. The "Glossary of Indian medicinal plants with active principles" yielded information of several alkaloids not discussed in ancient Materia Medica. Data collected from ancient Materia Medica coupled with internet search, provided missing information on several alkaloids.

For example, Materia Medica describes the presence of alkaloid in *Achyranthes aspera* L.N.O.—Amaranthaceae. Internet search provided information about name (achyranthine) and pharmacology of the alkaloid.[34,35]

Information on relatively lesser known alkaloids like abromine (Fig. 12.12), buchanine, christembine, gymnamine, marmeline, punarnavine and rohitukine (Fig. 12.13), a potential candidate with immunopharmacological activity, was obtained.[36-42]

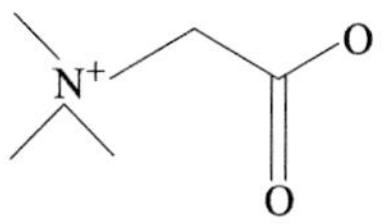

Figure 12.12 Structure of abromine.

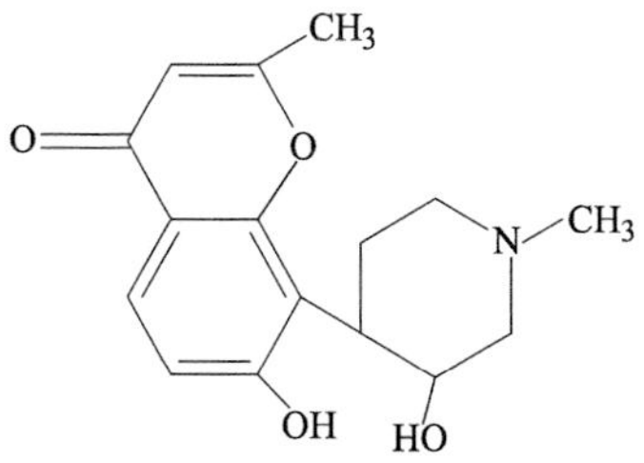

Figure 12.13 Structure of rohitukine.

Celatrus paniculatus Willd. N.O. Celastraceae, popularly known as *Jyotishmati* in Ayurveda is reported to contain two alkaloids, celastrine and paniculatine (Fig. 12.14).[43] The former was recorded to be powerful stimulant and pyrogenic. Loturine (Fig. 12.15), loturidine and colloturine, present in *Symplocos racemosa* Roxb. (Symplocaceae), are chemically related to harmine found in *Peganum harmala* L. (Rutaceae).[19] Data search on Amaryllidaceae yielded valuable information on alkaloids.[44-46]

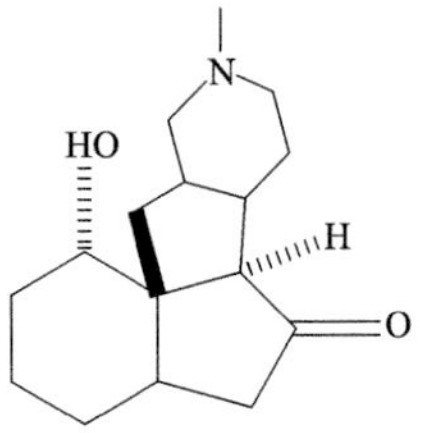

Figure 12.14 Structure of paniculatine.

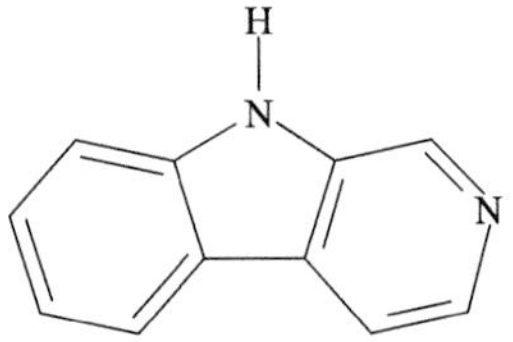

Figure 12.15 Structure of loturine.

Gentianine (Fig. 12.16) present in medicinal plants of Gentianaceae has been reported to be antipsychotic[47], antimalarial and anti-amoebic activities.[48] Phyllantidine and phyllantine (Fig. 12.17) were reported from leaves, fruits, and *in vitro* tissue cultures of *Emblica officinalis* Gaertn (Euphorbiaceae).[49]

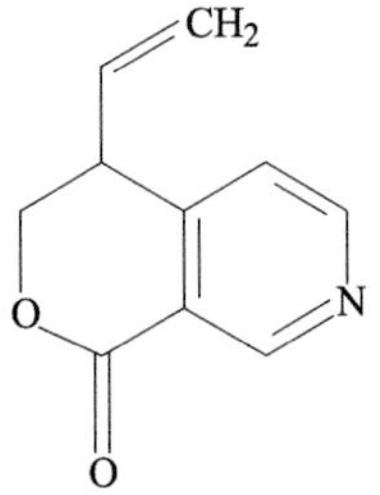

Figure 12.16 Structure of gentianine.

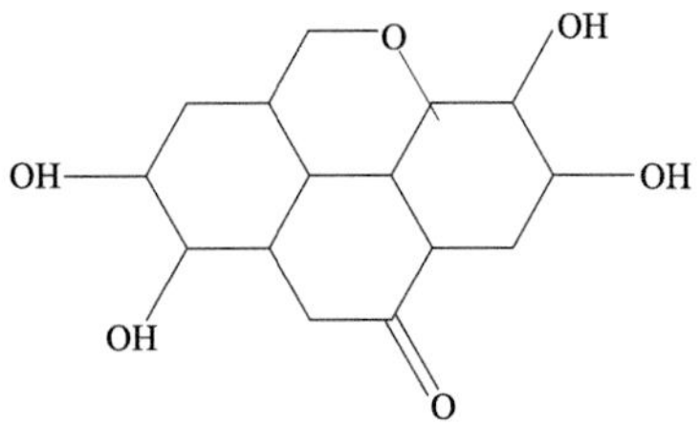

Figure 12.17 Structure of phyllantine.

Information on rare alkaloids such as cucurbitine (*Benincasa hispida*) (Thunb) Cogn. (Cucurbitaceae), pelosine and cissampeline (Fig. 12.18) (*Cissampelos pareira* Linn. (Menispermaceae), a muscle relaxant, hydrocotyline (*Centella* (*Hydrocotyle*) *asiatica* Linn. (Apiaceae), alpha pederine and beta pederine (*Paederia foetida* Linn. (Rubiaceae), ecliptine (*Eclipta alba* (L.) Hassk. (Asteraceae), echitamine (Fig. 12.19) and echitanine (*Alstonia scholaris* R.Br. (Apocynaceae), actinidine (Fig. 12.20) (*Nardostachys jatamansi* DC. (Valerianceae), herpestine and brahmine (*Bacopa monneira* Linn. (Scophulariaceae), anisotine, 3-hydroxyanisotine, vanestine, desmethoxyaniflorine and desmethoxyvascinone (*Adhatoda vasica* Nees. (Acanthaceae), cannabine (*Cannabis sativa* Linn. (Cannabaceae), calamine (*Acorus calamus* Linn. (Araceae), coptine, copsine, coptisine and palmatine (*Coptis teeta* Wall. (Ranunculaceae), delphinine and staphisagrine (*Delphinium denudatum* (Ranunculaceae), nigellidine (*Nigella sativa* L. (Ranunculaceae) was recorded during literature search.

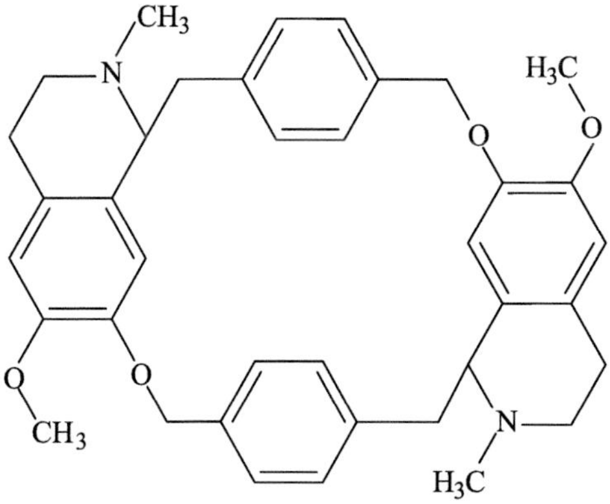

Figure 12.18 Structure of pelosine (cissampeline).

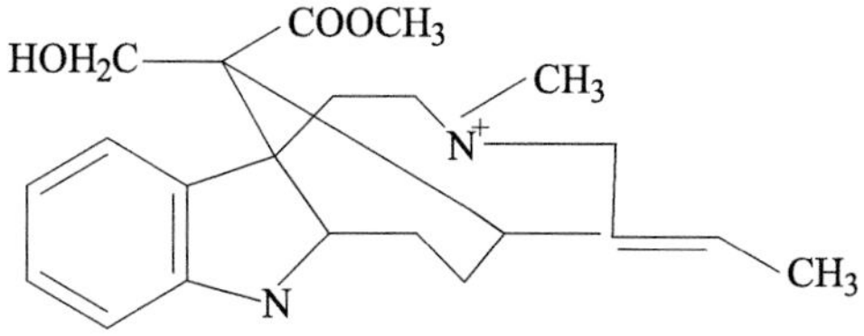

Figure 12.19 Structure of echitamine.

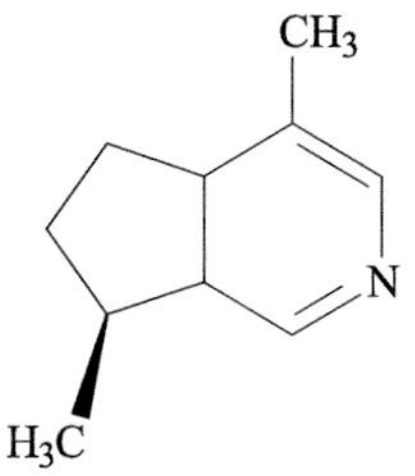

Figure 12.20 Structure of actinidine.

REFERENCES

1. Vaidya AD, Devasagayam TP. Current Status of Herbal Drugs in India: An Overview. *J Clin Biochem Nut* 2007; 41: 1-11.
2. Arora S, Kaur K, Kaur S. Indian medicinal plants as a reservoir of protective phytochemicals. *Teratogensis, Carcinogenesis, and Mutagenesis* 2003; 23: 295-300.
3. Singh AP, Malhotra S. A Review of Pharmacology of Phytochemicals from Indian Medicinal Plants. *Int J Alter Med* 2007; 5: 1.
4. Singh AP. Promising Phytochemicals from Indian Medicinal Plants. *Ethnobotanical Leaflets*, Southern Illinois University, 2005; Carbondale Edition.
5. Gupta KC, Chopra IC. A short note on anti-bacterial properties of chaksine: an alkaloid from *Cassia absus* Linn. *Indian J Med Res* 1953; 41: 459-460.
6. Chatterjee M, Dey CD. Pharmacological studies on chaksine chloride. IV. Action on some isolated tissues. *Bull Calcutta Sch Trop Med* 1962; 10: 113-114.
7. Cheema MA, Priddle OD. Pharmacological investigations of isochaksine, an alkaloid isolated from the seeds of *Cassia absus* Linn. (Chaksu). *Arch Int Pharmacodyn Ther* 1965; 158: 307-13.
8. Nadkarni AK, Nadkarni KM. *Indian Materia Medica*. 1976. Popular Prakshan, Mumbai, pp: 81.
9. Chakarborty A, Chowdhury BK, Bhattacharya P. Clausenol and Clausenine-two carbazole alkaloids from *Clausena anisata*. *Phytochemistry* 1995; 40: 295-298.
10. Bajad S, Bedi KL, Singla AK, *et al.* Antidiarrheal activity of piperine in mice, *Planta Medica* 2001; 67: 284-287.
11. Visen B, Shukla G, Patnaik S, *et al.* Hepatoprotective activity of *Ricinus communis* leaves. *Int J Pharmacogn* 1992; 30: 241-250.
12. Lalla JK, Hamrapurkar PD, Mukherjee SA. Isolation and characterization of Sausurine. *Indian Drugs* 2005; 42: 209-212.
13. Chopra IC. *Indigenous Drugs of India*. Academic Publishers. 1958, Kolkata, West Bengal, India.
14. No authors listed. Aegeline, an alkaloid of *Aegle marmelos. Ind J Med Res* 1968; 56: 237.
15. Gopalakrishnan C, Shankaranarayan D, Kameswaran L, *et al.* Pharmacological investigations of tylophorine, the major alkaloid of *Tylophora indica. Indian J Med Res* 1979; 69: 513-520.
16. Ghosal S, Amrithalingam S, Mukhopadhyay M, *et al.* Effect of lycoriside, an acylglucosyloxy alkaloid, on mast cells. *Pharm Res* 2004; 3: 240-243.
17. Kumaraguru Atul VS, Dhananjayan R. Effect of aegeline and lupeol-two cardio active principles isolated from leavers of *Aegele marmelos* Corr. *J Pharm Pharmacol* 1999; 51: 252.

18. Krishnamurthy TR. Some pharmacological actions of evolvine hydrochloride, *Curr Sci* 1959; 28: 64-65.
19. Chauhan NS. *Medicinal and Aromatic Plants of Himachal Pradesh.* 1999 Indus Publishing Company, New Delhi, pp: 59-62.
20. Sukumar E, Hamsaveni GR, Bhima RR. Pharmacological actions of cerpegin, a novel pyridine alkaloid from *Ceropegia juncea. Fitoterapia* 1996; 66: 403-406.
21. Sawant M, Isaac JC, Naraynan S. Analgesic studies on total alkaloids and alcohol extracts of *Eclipta alba* (Linn.) Hassk. *Phytother Res* 1997; 18: 111-113.
22. Barik BR, Bhaumik T, Patra A, *et al.* Premnazole, an isoxazole alkaloid of *Premna integrifolia* L. and *Gmelina arborea* L. with anti-inflammatory activity. *Fitoterapia* 1976; 639: 395.
23. Neogi NC, Rathor RS, Shrestha AD, *et al.* Studies on the anti-inflammatory and antiarthritic activity of achyranthine. *Ind J Pharmacol* 1969; 13: 37-47.
24. Ghosh MN, Singh H. Inhibitory effects of a pyrrolizidine alkaloid, crotalaburnine, on rat paw edema and cotton-pellet granuloma. *Brit J Pharmacol* 1974; 51: 503.
25. Dixit VP. Antifertility effects of solasodine (C27 H 43 0 2 N) obtained from *Solanum xanthocarpum* berries in male rats and dogs. *J Ster Biochem* 1986; 25: 27.
26. Saksena S, Dixit VK. Role of the total alkaloids of *Mucuna prurita* Baker in spermatogenesis in albino rats. *Indian J Natl Prod* 1976; 3: 3.
27. Ghosal S, Kumar Y, Singh S, *et al.* Chemical constituents of Amaryllidaceae. Part 21. Ungeremine and Criasbetaine, two anti-tumour alkaloids from *Crinum asiaticum. J Chem Research* 1986; 4: 112-113.
28. Ghosal S. Chemical constituents of Amaryllidaceae. Part 24. Crinafoline and Crinafolidine, two antitumour alkaloids from *Crinum latifolium. J Chem Research* 1986; 4: 312-313.
29. Rao KN, Bhattacharya RK, Veanlatchalam SR. Inhibition of thymidylate synthase by pergularinine, tylophorinidine and deoxytubulosine. *Indian J Biochem Biophys* 1999; 36: 442-448.
30. Rao KN, Bhattacharya RK, Veanlatchalam SR. Thymidylate synthase activity in leukocytes from patients with chronic muelocytic leukemia and acute lymphocytic leukemia and its inhibition by phenanthroindolizidine alkaloids pergularine and tylophorinidine. *Can Lett* 1997; 128: 183-188.
31. Padmaja R, Arun PC, Prashanth D, *et al.* Brine shrimp lethality bioassay of selected Indian medicinal plants. *Fitoterapia* 2002; 73: 508-510.
32. Badami S, Mnaohara Reddy SA, *et al.* Antitumor activity of total alkaloid fraction of *Solanum pseudocapsicum* leaves. *Phytother Res* 2003; 17: 1001-1004.
33. Dongre SH, Badami S, Ashok G, *et al. In vitro* cytotoxic properties of O-methylsolanocapsine isolated from *Solanum pseudocapsicum* leaves. *Indian J Pharmacol* 2007; 39: 208-209.

34. Neogi NC, Rathor RS, Shrestha AD, *et al.* Studies on the anti-inflammatory and antiarthritic activity of achyranthine. *Ind J Pharmacol* 1969; 3: 37-47.
35. Basu NK. The chemical constitution of achyranthine. *J Proc Inn Chem* 1957; 29: 73-76.
36. Dasgupta B, Basu K. Chemical investigation of *Abroma Augusta* Linn. Identity of abromine with betaine. *Experientia* 1970; 26: 477-478.
37. Datta SK, Sharma BN, Sharma PV. Buchanine, a novel pyridine alkaloid from *Cryptolepis buchanani. Phytochemistry* 1978; 17: 2047-2048.
38. Tyagi RD, Tyagi MK, Goyal HR, *et al.* A chemical Study on Krimiroga. *J Res Ind Med* 1982; 3: 130-132.
39. Rao GS, Sinsheimer JE, Mcllhenny HM. Structure of gymnamine, a trace alkaloid from *G. sylverstre* leaves. *Chem Ind* (London) 1965; 13: 537.
40. Sharma BR, Rattan RK, Sharma P. Marmeline, an alkaloid and other components of unripe fruits of *A. marmelos. Phytochemistry* 1981; 20: 2606-26071.
41. Agarwal RR, Dutt SS. Chemical examination of punarnava or *B. diffusa* Linn. 2. Isolation of an alkaloid punarnavine. *Chem Abst* 1936; 30: 3585.
42. Lakdawala AD, Shirole MV, Mandrekar SS, *et al.* Immunopharmacological potential of rohitukine: a novel compound isolated from the plant *Dysoxylum binectariferum. Asia Pac J Pharmacol* 1998; 32: 91-98, *Chem Abst* 109: 183182a.
43. Ahmad WS. A note on the chemical examination of *Celastrus paniculatus. Curr Sci* 1940; 9: 134-135.
44. Ghosal S, Saini KS, Razdan S, *et al.* Chemical constituents of Amaryllidaceae. Part 12. Crinasiatine, a novel alkaloid from *Crinum asiaticum. J Chem Res* 1985; 3: 100-101.
45. Ghosal S, Shanthy A, Kumar A, *et al.* Palmilycorine and lycoriside: Acyloxy and acylglucosyloxy alkaloids from *Crinum asiaticum. Phytochemistry* 1985; 24: 2703-2706.
46. Ghosal S. Chemical constituents of Amaryllidaceae. Part 24. Crinafoline and Crinafolidine, two antitumour alkaloids from *Crinum latifolium. J Chem Res* 1986; 4: 312-313.
47. Bhattacharya SK, Ghosal S, Chaudhuri RK, *et al.* Chemical constituents of gentianaceae XI: Antipsychotic activity of gentianine. *J Pharma Sci* 1976; 63: 1341-1342.
48. Natrajan PN, Wan AS, Zaman V. Antimalarial, antiamoebic and toxicity tests on gentianine. *Planta Med* 1976; 25: 258-260.
49. Khanna P, Bansal R. Phyllantidine & phyllantine from *Emblica officinalis* Gaertn leaves, fruits, & in vitro tissue cultures. *Indian J Exp Biol* 1975; 13: 82-83.

Chapter 13

Scientifically Validated and Novel Anti-arthritic Ayurvedic Drugs

13.1 INTRODUCTION

Arthritic diseases have been known to exist since antiquity. Probably the earliest description of arthritis occurs in *Athravaveda*, around 1000 B.C.[1] *Charaka Samhita*, written in the post Vedic period, has dealt more accurately with the etiology, symptomatology, diagnosis and treatment of arthritis. Prognosis of arthritis, as proclaimed by the ancient physicians of India, remains unaltered.[2]

13.2 PATHOLOGY OF ARTHRITIS IN AYURVEDA

Ayurveda suggests that arthritis is caused primarily by an excess of *ama* (byproduct of improper digestion) and lack of *agni* (digestive fire). This can be caused by poor digestion and a weakened colon, resulting in the accumulation of undigested food and the buildup of waste matter. Poor digestion allows toxins to accumulate in the body, and problems with the colon allow the toxins to reach the joints.[3]

According to Ayurveda in order to treat arthritis, the agni needs to be stimulated and the ama suppressed. Ayurveda distinguishes three categories of arthritis, corresponding to vata, pitta and kapha. The clinical features of these three types of arthritis are enumerated in the Table 13.1.

Table 13.1 Clinical features of three types of arthritis according to Ayurveda

Type	*Features*
Vata	Joints crack, pop and become dry, but are not swollen.
Pitta	This arthritis is characterized by inflammation. The joint becomes swollen, is painful even without movement. It often looks red and feels hot to the touch.
Kapha	In kapha-type arthritis, the joint also becomes stiff and swollen, but it feels cold and clammy rather than hot. A little movement, rather than aggravating the pain, tends to relieve it.

13.3 CLASSIFICATION OF ARTHRITIS IN AYURVEDA[4-7]

1. *Aamvata* (Rheumatoid arthritis): In Ayurveda there is a very detailed description about rheumatoid arthritis or "Aama vata". According to ayurvedic pathology, amvata is caused by ama, a toxin that is produced by imbalanced body fire). The toxin ama is carried by imbalanced vata and reaches the kapha-dominated joints.
2. *Vatrakta* (Gout): Vatarakta is a condition in which, the blood and vata dosha both get vitiated and troubles an individual. It is compared to gout which refers to a group of metabolic disorders. In these disorders, the crystals of sodium urate get deposited in the tissues of the body. This results in the raised levels of uric acid in the blood. The typical gout like condition catches the small joints of the body first.
3. *Sandhivata (Osteroarthritis)*: *Sandhivata, sandhi* means the joint and *vata,* the vata dosha. When activities of the *vata* increase inside the *sandhis* or joints, it is known as the *Sandhivata. Vata* is dry in nature so it absorbs the fluidity, from any part of the body and is also destructive or catabolic in nature, due to these two reasons *vata* causes destruction of the cartilage and reduction in the synovial fluid inside the joint capsule.

13.4 AYURVEDIC FORMULATIONS FOR ARTHRITIS[8-21]

Table 13.2 Classical Ayurvedic formulations for Arthritis

S.No	*Formulation*	*Disease*	*Reference*
1.	Shuddha Guggulu (*Commiphora mukul*)	Rheumatoid arthritis	Mahesh et al., 1981
2.	Bhallataka (*Semecarpus anacardium*)	Rheumatoid arthritis	Upadhyay et al., 1986
3.	Gourakh (*Dalbergia lanceolaria*)	Rheumatism	Tripathi and Shukla, 1964
4.	Simhanada guggulu with guduchi and sunthi	Rheumatoid arthritis	Babu, 1982

5.	Eranda (*Ricinus communis*)	Rheumatoid arthritis	Shastri, 1981
6.	Rasnadiguggulu	Rheumatoid arthritis	Shukla et al., 1985
7.	Guduchi and sunthi	Rheumatoid arthritis	Sastry, 1977
8.	Kanchanara gugulu kwatha	Rheumatism	Rao, 1982
9.	Vatahari guggul	Rheumatism	Pandey et al., 1986
10.	Shambalu (*Vitex negundo*)	Rheumatoid arthritis	Mohiddin, 1979
11.	Amber (Ambra grasea)	Rheumatoid arthritis	Mutharasan, 1977
12.	Cheriya rasnadi kashayam	Anti-arthritic	Valarmati et al., 2004
13.	Rasonadi kvatha	Anti-rheumatic	Kishore and Banerjee, 1973
14.	Vachadi guggulu and Vyoshadi guggulu	Anti-rheumatic	Chandrasekhara, 1966

13.5 SCIENTIFIC RESEARCH ON CLASSICAL ANTI-RHEUMATIC AYURVEDIC FORMULATIONS

Vyas and Shukla studied the efficacy of purified guggul in 35 patients of rheumatoid arthritis in order to assess its anti-rheumatic activity, dose requirement, resistance development, side effects and effects on hematology. The results indicated that guggul acts as an anti-rheumatic agent without toxic or side effects.[22]

A clinical study by Das evaluated the efficacy of nirgundi taila in the management of osteoarthritis. The researchers found that nirgundi taila was much more effective than the control drug (diclofenac sodium), though maximum improvement was reached after a longer time of treatment.[23]

Singh et al., investigated efficacy of *Commiphora mukul* in two studies. In a first experimental case study, *C. mukul* was reported to be beneficial in the treatment of osteoarthritis.[24] The success of the previous study outcome, led to the quasi-experimental model being used for investigating the further use of *C. mukul* in osteoarthritis. *C. mukul* was administered to 30 male and female participants in a capsule form (500 mg concentrated exact delivered TID) along with food. At the end of treatment, there was a significant difference in the scores of the primary and secondary outcome measures.[25]

Chopra et al., in a 32-week randomized, placebo-controlled clinical evaluation studied the efficacy of RA-11 (*Withania somnifera, Boswellia serrata, Zingiber officinale,* and *Curcuma longa*) on osteoarthritis in the knees in 358 patients. Compared with a placebo, RA-11 demonstrated potential efficacy and safety in the symptomatic treatment of osteoarthritis knees over 32 weeks of therapy.[26]

Park and Ernst systematically reviewed all randomized controlled trials on the effectiveness of Ayurvedic medicine for rheumatoid arthritis. Reputed databases were explored for generating data regarding randomized controlled trials of Ayurvedic medicine for rheumatoid arthritis. The authors concluded that the existing randomized controlled trials failed to show convincingly that such treatments are effective therapeutic options for rheumatoid arthritis.[27]

Sumantran et al., assessed the chondroprotective potential of Triphala guggulu and Triphala shodith guggulu by examining its effects on the activities of pure hyaluronidase and collagenase type 2 enzymes. Aqueous and hydro-alcoholic extracts of Triphala shodith guggulu showed weak but dose-dependent inhibition of hyaluronidase activity. In contrast, the Triphala guggulu was 50 times more potent than the Triphala shodith guggulu extract with respect to hyaluronidase inhibitory activity.[28,29]

Furst et al., in a short communication highlighted the possibility of carrying out well-controlled, double-blind, placebo-controlled trials of classical Ayurvedic treatment. A total of 46 patients ≥ 18 yr of age, with active rheumatoid arthritis, were randomized to one of three outpatient treatment groups: methotrexate + placebo Ayurveda; Ayurveda + placebo methotrexate; and methotrexate + Ayurveda. The traditional dosage forms were used in the study. Six placebos were also formulated, with each placebo representing a dosage form.[30]

Falkenbach and Oberguggenberger documented the use of Ayurveda in ankylosing spondylitis and low back pain. According to Ayurveda, rheumatic symptoms result from an inequality and disharmony among the three biological humors, in particular from a predominance and dysfunction of vata. Based on this principle, treatment of ankylosing spondylitis and low back pain falls within the purview of Ayurveda.[31]

13.6 RECENT ADVANCES IN ANTI-RHEUMATIC DRUG RESEARCH FROM HERBAL DRUGS

13.6.1 *Ammania baccifera* Linn. (Lythraceae)

The alcoholic and aqueous extracts of *A. baccifera* significantly ($P < 0.01$) decrease the paw edema on the 28th day in albino rats in chronic inflammatory models like cotton pallet induced granuloma and complete Freund's Adjuvant. In both the inflammatory models, alcoholic extracts show more potency than the aqueous extracts in terms of percentage of inhibition of inflammation.[32]

13.6.2 *Alangium salviifolium* Wang (Alangiaceae)

Petroleum ether, ethyl acetate, chloroform and methanol extracts of *A. salviifolium* exhibited significant anti-arthritic activity in Fruends adjuvant arthritis model. The present study suggests that further investigations are needed in exploring anti-arthritic of *A. salviifolium*.[33]

13.6.3 *Merremia tridentata* (L.) Hall. f. (Convolvulaceae)

The ethanol extract of *M. tridentata* exhibited significant dose-dependent activity in acute inflammation and the doses of 100 mg/kg body weight and 200 mg/kg body weight produced 38.3 and 42.8 percent inhibition respectively after 3 hr as compared with that of the standard drug which showed 48.5 percent inhibition. In the arthritis model, the doses of 100 mg/kg body weight and 200 mg/kg body weight of the ethanol extract produced 49.0 and 51.7 percent inhibition respectively after 19 days when compared with that of the standard drug (55.5 percent).[34]

13.6.4 *Premna serratifolia* Linn. (Verbenaceae)

In Freund's adjuvant induced arthritis model, the ethanol extract *P. serratifolia* at the dose of 300 mg/kg body weight inhibited the rat paw edema by 68.32 percent which is comparable with standard drug indomethacin 74.87 percent inhibition of rat paw edema after 21 days.[35]

13.6.5 *Strobilanthus callosus* Nees and *Strobilanthus ixiocephala* Benth. (Acanthaceae)

Lupeol (in the doses of 200, 400 and 800 mg/kg) and 19α-H-lupeol isolated from the roots of *S. callosus and S. ixiocephala* respectively, produced a dose dependent inhibition *i.e.* 24, 40 and 72 percent where as 19α-H-lupeol showed 21, 47 and 62 percent inhibition after 24 hr in the acute model of inflammation. In the chronic model of granuloma pouch in rats, lupeol exhibited 33 percent and 19α-H-lupeol, 38 percent reduction in granuloma weight. In the arthritis model, lupeol exhibited 29 percent and 19α-H-lupeol 33 percent inhibition after 21 days respectively.[36]

13.6.6 *Barringtonia racemosa* (L.) Spreng. (Lecythidaceae)

The ethyl acetate fraction of *B. racemosa* displayed potent anti-inflammatory activity in Complete Freund's Adjuvant induced arthritis in rats. Activity guided fractionation led to isolation of bartogenic acid, which was evaluated for effectiveness against Complete Freund's Adjuvant-induced arthritis in rats. Bartogenic acid at doses of 2, 5 and 10 mg/kg/day, PO (by mouth), protects rats against the primary and secondary arthritic lesions, body weight changes and hematological perturbations induced by Complete Freund's Adjuvant.[37]

13.6.7 *Gymnema sylvestre* R.Br. (Asclepiadaceae)

The petroleum ether and aqueous extract of leaves of *G. sylvestre* were studied for anti-arthritic activity in Freund's adjuvant induced arthritis in rats. The study revealed that both extracts of *G. sylvestre* possessed significant anti-arthritic activity in all parameters of the study compared to the control group.[38]

13.6.8 *Moringa oleifera* Lam. (Moringaceae)

A hydroalcoholic extract of *M. oleifera* flowers decreased paw edema volume (primary lesion), inflammation at non-injected sites of left hind paw, and arthritic index (secondary lesion) in diseased animals as compared with untreated control animals. Further, it decreased the serum levels of Rheumatoid Factor and levels of the cytokines tumor necrosis factor-α and interleukin-1.[39]

13.6.9 *Justicia gendarussa* Burm. f. (Acanthaceae) and *Withania somnifera* Dunal (Solanaceae)

Ethanolic extracts of *J. gendarussa* and *W. somnifera* suppressed the anti-arthritic changes induced in male albino rats using Freund's complete adjuvant and bovine type II collagen. Both of the drugs produced statistically significant results.[40]

13.6.10 *Snake Venom*

In Ayurveda, cobra venom was used to treat joint pain, inflammation and arthritis. The *visachikitsha*, the division of Ayurveda deals with the use of venoms to cure diseases through the Ayurvedic technique known as *suchikavoron* (venom at the tip of a needle) and *shodhono* (detoxification of venom) was able to treat several chronic diseases.[41]

A study confirmed that the Indian monocellate cobra (*Naja kaouthia*) venom significantly antagonized the changes in the arthritis biomarkers in an experimental animal model, where arthritis was induced by Freund's complete adjuvant.[42]

REFERENCES

1. Srurrock MD, Sharma JN, Buchanan MD. Evidence of rheumatoid arthritis in ancient India. *Arthritis & Rheumatism* 2005; 20: 42-44.
2. Sharma JN, Sharma JN. Arthritis in ancient Indian literature. *Bull Deptt Pharmacol AIIMS*, 1973; 8: 37-42.

3. Dubey GP, Singh RH. A preliminary study on certain psychosomatic factors in cases of different types of arthritis. *Rheumatism,* 1967; 2: 133-144.
4. Shanbagh VV. *Aamvaat* (rheumatoid arthritis). *Deerghayu Int* 6: 11-12.
5. Narayanaswami, V. Rheumatoid arthritis (*amavata*). *Nagarjun* 1978; 21: 18-19.
6. Mishra LC. *Rheumatoid Arthritis, Osteoarthritis, and Gout.* Lakshmi Chandra Mishra (Ed.), 2004; 167-183.
7. Anonymous. *The Ayurvedic pharmacopoeia of India,* Govt. of India. Ministry of Health and Family Welfare, Department, I.S.M.Q.H. Ist ed. Part I, Vol. 3; 2001
8. Mahesh S, Pandit M, Hakala C. A study of shuddha guggulu on rheumatoid arthritis. *Rheumatism* 1981; 16: 54-67.
9. Upadhyay BN, Singh TN, Tewari CM, *et al.* Experimental and clinical evaluation of *Semecarpus anacardium* nut (*bhallataka*) in the treatment of *amavata* (rheumatoid arthritis). *Rheumatism* 1986; 21: 22-24.
10. Tripathi SN, Shukla KP. Rheumatism (*amavata*) and its treatment with an indigenous drug, gourakh (*Dalbergia lanceolaria*). *J Med Sci* 1964; 5: 75-89.
11. Babu SR. Effect of *simhanada guggulu* with *guduchi* and *sunthi* in *amavata. Rheumatism* 1982; 18: 11-13.
12. Shastri MS. Effects of *eranda* in *amavata. Rheumatism* 1981; 16: 149-152.
13. Shukla KP, Singh SP, Kishore N, *et al.* Evaluation of *rasnadiguggulu* compound in the treatment of rheumatoid arthritis. *Rheumatism* 1985; 21: 1-5.
14. Sastry MM. Clinical trial of *guduchi* and *sunti* on *amavata. Rheumatism* 1977; 2: 11-13.
15. Rao NH. Kanchanara *gugulu kwatha* in rheumatic diseases. *Rheumatism,* 1982; 2: 23.
16. Pandey VK. Evaluation of *vatahari guggul* and *nadivaspa sweda* in the management of rheumatic diseases. *Rheumatism* 1986; 22: 13.
17. Mohiddin SG. The role of *shambalu* (*Vitex negundo*) in rheumatoid arthritis. *Rheumatism* 1979; 14: 97-112.
18. Mutharasan S. *Amber* (Ambra grasea) *mezhugu* for rheumatoid arthritis. *Rheumatism* 1977; 12: 38-43.
19. Valarmati RS, Sundari KK, Ramya S, *et al.* Anti arthritic activity of *cheriya rasnadi kashayam. Aryavaidyan* 2004; 18: 49-50.
20. Kishore P, Banerjee SN. Clinical evaluation of *rasonadi kvatha* in the treatment of amavata- rheumatoid arthritis. *J Res Ayur Sidd* 1995; 9: 29-37.
21. Chandrasekhara HI. Clinical trial on rheumatoid arthritis with *Vyoshadi guggulu* and *Vachadi guggulu, Rheumatism* 1982; 17: 127-130.
22. Vyas SN, Shukla CP. A clinical study on the effect of *shudha guggulu* in rheumatoid arthritis. *Rheumatism* 1987; 23: 15-26.

23. Das B, Padhi MM, Singh OP, *et al.* Clinical evaluation of nirgundi taila in the management of sandhivata. *Ancient Sci Life* 2001; 23: 22-34.
24. Singh BB, Mishra L, Aquilina N, *et al.* Usefulness of guggul (Commiphora mukul) for osteoarthritis of the knee: an experimental case study. *Altern Ther Health Med* 2002; 120: 112-114.
25. Singh BB, Mishra L, Aquilina N, *et al.* The effectiveness of *Commiphora mukul* for osteoarthritis of the knee: an outcomes study. *Altern Ther Health Med* 2003; 9: 74-79.
26. Chopra A, Lavin P, Patwardhan B, *et al.* A 32-week randomized, placebo-controlled clinical evaluation of RA-11, an ayurvedic drug, on osteoarthritis of the knees. *J Clin Rheumatol* 2004; 10: 236-245.
27. Park J, Ernst E. Ayurvedic medicine for rheumatoid arthritis: a systematic review. *Sem Arthritis Rheum* 2005; 34: 705-713.
28. Sumantran VN, Kulkarni AA, Harsulkar A, *et al.* Hyaluronidase and collagenase inhibitory activities of the herbal formulation *triphala guggulu*. *J Biosci* 2007; 32: 755-761.
29. Furst DE, Venkatraman MM. Well controlled, double-blind, placebo-controlled trials of classical ayurvedic treatment are possible in rheumatoid arthritis. *Ann Rheum Dis* 2011; 70: 392-393.
30. Falkenbach A, Oberguggenberger R. Ayurveda in ankylosing spondylitis and low back pain. *Ann Rheum Dis* 2003; 62: 276-277.
31. Tripathy S, Pradhan D, Anjana M. Anti-inflammatory and antiarthritic potential potential of *Ammania baccifera* Linn. *Int J Pharm Bio Sci* 2010; 1: 1-7.
32. Jubie S, Jawahar N. Antiarthritic activity of bark extracts of *Alangium salviifolium* Wang. *Rasayan J Chem* 2008; 1: 433-436.
33. Kamalutheen M, Gopalakrishnan S, Ismail TS. Anti-inflammatory and anti-arthritic activities of *Merremia tridentata* (L.) Hall. f. *El J Chem*, 2009; 6: 943-948.
34. Rajendran R, Krishnakumar E. Anti-arthritic activity of *Premna serratifolia* Linn., wood against adjuvant induced arthritis. *Avicenna J Med Biotech*, 2010; 2: 101-106.
35. Agarwal RB, Rangari V. Anti-inflammatory and antiarthritic activities of lupeol and 19α-H lupeol isolated from *Strobilanthus callosus* and *Strobilanthus ixiocephala* roots. *Ind J Pharmacol* 2003; 35: 384-387.
36. Patil KR, Patil CR, Jadhav RB, *et al.* Anti-arthritic activity of bartogenic acid isolated from fruits of *Barringtonia racemosa* roxb. (Lecythidaceae). eCAM Advance Access published online on September 21, 2009. *eCAM*, doi:10.1093/ecam/nep148.
37. Malik JK, Manvi FV, Nanjware BR, *et al.* Anti-arthritic activity of leaves of *Gymnema sylvestre* R.Br. leaves in rats. *Der Pharmacia Lettre* 2010; 2: 336-341.
38. Mahajan, SG, Mehta AA. Anti-arthritic activity of hydroalcoholic extract of flowers of *Moringa oleifera* lam. in wistar rats. *J Herbs Spices Med Plants* 2009; 15: 149-163.

39. Paval J, Kaithrti SK, Potu BK, *et al.* Comparing the anti-arthritic activities of the plants *Justicia gendarussa* Burm F. and *Withania somnifera* Linn. *Int J Green Pharm* 2009; 3: 281-284.
40. Debnath PK, Chaturvedi GN, Bhattacharya SK, *et al.* Competitive study of some pharmacological actions of crude and shodhita cobra venom. *J Res Ind Med* 1972; 7: 54.
41. Gomes A. Snake Venom - An Anti Arthritis Natural Product. *Al Ameen J Med Sci* 2010; 3: 176.
42. DeJager P. Turmeric, the ayurvedic spice of life., *LOAJ* 2003; 1: 3: 13.

Chapter 14

Rajpatha-Ethno Medicine of Controversial Origin

14.1 INTRODUCTION

Many plant drugs documented in Ayurvedic textbooks are controversial on their accurate botanical linkages.[1] Since plants and plant drugs in Ayurveda were designated Sanskrit names, often based on the "doctrine of signature", morphological appearance, properties and action, the interpretation of these names during the later period of time led to acceptance of more than one botanical species for one plant drug.[2]

14.2 VARIETIES OF *PATHA*

Two varieties of *patha* have been mentioned in Ayurvedic texts, viz. *brhat patha* (*rajpatha*) and *laghu patha*, i.e., with large and small leaves, respectively. Both the varieties are, more or less, similar in their properties.[3] *Laghu patha* has been identified as *Cissampelos pareira* Linn. *Brhat patha* has been mentioned as *Kuchelika* in *Kaydeva Nighantu*.[4] *Charaka* has also indicated *Patha* and *Kuchela* as separate drugs in Shakvarga.[5,6]

In Ayurvedic Materia Medica, *Cycela peltata* Lam. and *Stephania hernandifolia* (Willd) Walp. are taken as *brahat patha*. *Cissampelos pareira*, *C. peltata* and *S. hernandifolia* are members of family Menispermaceae.[7] As per Ayurvedic herbology, *brhat patha* is bitter, astringent and beneficial in blood-borne diseases and polyuria.[8] Some authors have indicated *C. peltata* as a possible substitute for *C. pareira*.[9]

14.3 DISTINCTION BETWEEN *CYCELA PELTATA* AND *STEPHANIA HERNANDIFOLIA*

The roots of *C. pareira* can be distinguished from *C. peltata* and *S. hernandifolia* by the presence of high concentration of pharmacologically active alkaloid bebeerine, which was found to be present in very low concentration in *Stephania japonica* and absent in roots of *Cyclea peltata*. The roots of *Cyclea peltata* were found to contain a high concentration of saponins and comparatively in low concentration in *Cissampelos pareira* where it was found to be absent in roots of *Stephania japonica*.[9]

14.4 ETHNOPHARMACOLOGY OF *CYCELA PELTATA* AUCT. NON (LAMK) HOOK. F. & THOMSON. & THOMSON.

Syn: *C. barbata*

14.4.1 Distribution and Botany

C. peltata Lam. grows throughout India and Sri Lanka, up to 800-900 m elevation. It is commonly known as Green grass jelly. The plant is a slender twining shrub, frequently climbing on tall trees. The leaves are simple, alternate, heart shaped, 2.5-10 cm long and 2.5-3.75 cm broad, stipule 5-10 cm long and nerves 7-11. The flowers unisexual, pale yellow, in axillary panicles. The fruits are ovoid drupes, brown or scarlet in color. The seeds are covered. The roots are tuberous, cylindrical, irregularly curved, with a grayish brown surface. The plant blooms in the rainy season.[9]

14.4.2 Phytochemistry

The roots of *C. peltata* are reported to contain alkaloids including *d*-tetrandrine, *dl*-tetrandrine, *d*-isochondrodendrine, fangchinoline, tetrandrine N-2′-oxide, *α*-cyclanoline, tetrandrine 2′*β*-N-oxide,(–)-2-norlimacine, (–)-curine, (–)-cycleapeltine, (–)-N-methylcoclaurine, (–)-repandine, (+)-coclaurine, (+)-cycleabarbatine, (+)-cycleanorine, coclaurine and cycleadrine.[10-13]

14.4.3 Traditional Medicinal Use

In tribal medicine, jelly sediment obtained from *C. peltata* is applied to the head and washed after 30 min in Kerala.[14] The roots are used for curing coryza, hemorrhoids, diarrhea and a burning sensation in Mangalore, in the state of Karnataka, India.[15]

14.4.4 Pharmacological Research

Alkaloids isolated from petroleum ether and methanolic extracts alkaloids and their methiodides showed activity similar to *d*-tubocurarine.[10] Bisbenzylisoquinoline alkaloids isolated from *C. peltata* have demonstrated cytotoxic and antimalarial activities.[16]

In ethylene glycol treated animals, simultaneous administration of the powdered root of *C. peltata* resulted in decreased urinary oxalate and calcium. Likewise, serum potassium was lowered and magnesium was elevated.[17]

In an investigatory study, 70 percent methanolic leaf extract of *C. peltata* significantly changed the increased malonyldyaldehyde level and decreased glutathione levels found in rats treated with cisplatin alone.[18] Pretreatment with *C. peltata* extract provided significant protection against the peptic ulceration caused by ethanol administered individually, or in combination with indomethacin.[19]

14.5 ETHNOPHARMACOLOGY OF *STEPHANIA HERNANDIFOLIA* (WILLD) WALP.

Syn: *S. japonica*

14.5.1 Botany

S. hernandifolia is a woody smooth vine. The leaves are oval or sub-rounded-oval in shape, 6 to 15 cm in length, and 4 to 12 cm in width, with obtuse and nearly retuse apex and rounded base, and smooth on both surfaces, with long petioles (4 to 12 cm long). The inflorescences are in umbels on peduncles 3 to 4 cm in length. The male and female flowers are small and pale yellow. The fruit is red, small, rounded but flattened, about 8 mm long and 6 mm wide.[20]

14.5.2 Phytochemistry

Bancroft in a study noted that the extract of the roots was exceedingly poisonous for frogs. The physiological action of the active constituent appeared to be identical with that of *picrotoxin*, the active principle of *Cocculus*, a genus of the same order as *Stephania*. Bancroft failed to obtain *picrotoxin* from the plant, and thus suspected the poisonous effects to be due to an alkaloid.[21] Investigative work on the plant in 1924 reported alkaloids, metastephanine, stephanine and protosetaphine along with a phenol base and base.[22]

The plant contains alkaloids including hernandifoline and hernandiline.[23] The roots of *S. hernandifolia* from Mangalore yielded *d*-tetrandrine, fangchinoline, *d*-tetrandrine, and *d*-isochondrodendrine.[24] Epistephanine, (+)-3′,4′-Dihydrostephasubine and methylhernandine have been reported.[25-27]

Chief alkaloids of *S. hernandifolia* are reported to be antispasmodic.[28] The bulb of *S. hernandifolia*, used by the local people and traditional healers in the eastern Himalayan belt, were studied for their effects on serum glucose levels in non-diabetic and diabetic rat models at different prandial states. *S. hernandifolia* increased the serum glucose levels of non-diabetic rats in all the series of experiments ($P < 0.05$ or $P < 0.01$). In NIDDM model rats, *S. hernandifolia* had a tendency to raise the serum glucose level.[29]

Several studies have reported effect of *S. hernandifolia* on testicular activity in rats. In one study, adult male Wistar rats, were forcefully fed with the aqueous extract of these leaves at the dose of 2 g or 4 g of leaves/ ml distilled water/100 g body weight/day for 28 days. Treatment with this leaf extract at both doses resulted in significant reduction in relative weight of the sex-organs. Further the treatment with the extract resulted in diminution in the activity plasma level of testosterone along with inhibition of spermatogenesis without any induction of hepatic and renal toxicity.[30]

In another study, the testicular inhibitory effect of the aqueous fraction of methanol extract of *S. hernandifolia* leaf was studied in male Wistar rats. The supernatent and the precipitate part of aqueous fractions of the methanol extract of the leaf were gavaged separately to rats at a similar dose of 200 mg/ml per 100 g body weight per day for 28 days. In both treated groups, there were significant decreases in the relative weights of the sex organs, the testicular key androgenic enzymes activities, the plasma level of testosterone, the number of different germ cells at stage VII of seminiferous epithelial cell cycle and the seminiferous tubular diameter in comparison to the controls.[31]

An ethno medicinal formulation (based on *S. hernandifolia)* at 500 and 250 mg/kg doses induced 66.7 and 33.3 percent post-coital pregnancy interception respectively, in Wistar rats. The higher dose exhibited significant reduction in number of litters born and also the anti-implantation property. In contrast, none of the dose levels of aqueous extract of *S. hernandifolia* interfered in pregnancy but a significant anti-implantation property was observed at doses of 2 and 1 g/kg, even as the higher dose produced significant reduction in number of litters born as well. HPTLC and HPLC analysis of both exhibited marked chemical differences.[32]

The n-hexane fraction of the hydroethanolic (1:1) extracts *of S. hernandifolia* leaves and *Achyranthes aspera* roots (in a composite manner at a ratio of 1:3, respectively), exhibited maximum spermicidal activity in human and rat spermatozoa. At a concentration of 0.1 g/ml hexane fraction, the entire sperm of the human sample were immobilized immediately (within 20 s). In case of the rat sample, all epididymal spermatozoa were immobilized immediately (within 20 s) by treatment with hexane fraction at a concentration of 0.004 g/ml. All human sperm were found to be nonviable within 20 min.[33]

14.5.3 Conclusions

Medicinal plants in Ayurveda were designated Sanskrit names, often based on the "doctrine of signature", morphological appearance, properties and action. Due to lack of correct identification, similar looking plants are often collected from the field site along with the genuine medicinal plant by mistake. Hence further studies leading to identification for Rajpatha are greatly warranted.

REFERENCES

1. Uniyal MR, Joshi GC. Historical view of the basic principles of the identification of controversial drugs, problems and suggestions. *Sach Ayur* 1993; 45: 531-536.
2. Dobriyal RM, Ananthanarayana DB. Controversial Nomenclature of Ayurvedic drugs: Challenges for Scientists. *Pharmacog Rev* 2009; 3: 1-7.
3. Bapalal V. *Nighantu Adarsh* (Purvardh) Shri Swami Atmanand Saraswati Ayurvedic Sanskrit Pharmacy Ltd. Surat, 1952, pp. 19.
4. Anonymous. *Kaydeva Nighantu*. Commentary by Sharma GP Chakumbha Orientalia, Varanasi, 2006, pp, 124-125.
5. Anonymous. *Charaka Samhita*. Commentary by Shastry, K.N. Chaukhambha Vidvybhavan, Varnasi 1970, pp.233.
6. Manilal KS, Sabu. T. *Cyclea barbata* Miers (Menispermaceae): a new record of a medicinal plant from South India. *Ancient Sci Life* 1985; 4: 229-231.
7. Nadkarni AK. *Indian Materia Medica*. Popular Prakashan, Bombay, 1976; pp, 14.
8. Khory R. *Materia Medica of India and Their Therapeutics*. Neeraj Publishers. 1927; pp. 25.
9. Prasad NBR, Girija Devi RS, Hepsibah PTA. *Cyclea peltata* Diels—A possible substitute for *Cissampelos pareira* Linn. *Ancient Sci Life* 1995; 15: 150-152.
10. Hullatti KK, Sharada MS. Comparative phytochemical investigation of the sources of ayurvedic drug Patha: a chromatographic fingerprinting analysis. *Ind J Pharm Sci* 2010; 72: 39-45.
11. Nadkarni KM. Indian Materia Medica. Vol. I, 3rd ed., Popular Book Depot, Bombay. Dhootapapaeshwar Prakashan Ltd. Panvel, 1927; pp. 219.
12. Kupchan SM, Yokoyama N, Thyagarajan BS. Menispermaceae alkaloids II. The alkaloids of *Cyclea peltata* Diels. *J Pharm Sci* 1953; 50: 164-167.
13. Kupchan SM, Liepa AJ, Baxter RL, *et al*. Tumor Inhibitors. LXXIX. New alkaloids and related artifacts from *Cyclea peltata*. *Org Chem* 1973; 38: 1846-1852.

14. Harborne G. *Phytochemical Dictionary, 2nd ed.*, Taylor and Francis, 1999; pp. 21.
15. Guinaudeau H, Lin LZ, Ruangrungsi N, *et al.* Bisbenzylisoquinoline alkaloids from *Cyclea barbata. J Nat Prod* 1993; 56: 1989-1992.
16. Vijayan A, Liju VB, Reena John JV, *et al.* Traditional remedies of Kani tribes of Kootor reserve forests, Agasthyavanam, Thiruvananthapuram, Kerala. *Ind J Trad Knowl* 2006; 6: 589-594.
17. Shiddamallayya N, Yasmeen A, Gopalkumar K. Medico-botanical survey of *Kumar parvatta* Kukke Subramanya, Manglore, Karnataka. *Ind J Trad Knowl* 2010; 9: 96-99.
18. Lin LZ, Shieh HL, Angerhofer CK, *et al.* Cytotoxic and antimalarial bisbenzylisoquinoline alkaloids from *Cyclea barbata. J Nat Prod* 1993; 56: 22-29.
19. Christina AJM, Packia Lakshmi M, Nagarajan M, *et al.* Modulatory effect of *Cyclea peltata* Lam. on stone formation induced by ethylene glycol treatment in rats. *Meth Find Exp Clin Pharmacol* 2002; 24: 77.
20. Vijayan FP, Rani VKJ, Vineesh VR, *et al.* Protective effect of *Cyclea peltata* Lam on cisplatin-induced nephrotoxicity and oxidative damage. *J Basic Clin Physiol Pharmacol* 2007; 18: 101-114.
21. Shinde VJ, Latha PG, Shyamal S, *et al.* Gastric antisecretory and antiulcer activities of *Cyclea peltata* (Lam.) Hook. f. & Thoms. in rats. *J Ethnopharmacol* 2009; 125: 350-355.
22. Prasad S, Wahi AK. Pharmacognostical studies on *Stephania hernandifolia* (Willd) Walp. (rajpatha). *J Res Ind Med* 1970; 5: 84-94.
23. Bancroft TL. On the poisonous constituents of *Stephania hernandifolia.* Proceedings of the Linnæan Society of New South Wales, 1889.
24. Kondo G, Sanada P. Alkaloids of *Stephania hernandifolia. J Pharm* 1924: 514: 163.
25. Fesenko DA, Fadeeva II, Il'inskaya, *et al.* Alkaloids of *Stephania hernandifolia* VI. Hernandifoline. *Chem Nat Comp.* 1971; 7: 122.
26. Kupchan SM, Wady L, Asbun BS, *et al.* Menispermaceae alkaloids III. Alkaloids of *Stephania hernandifolia. J Pharm Sci* 1961; 50: 819-822.
27. Ray AB, Tripathi RM, Gambhir SS, *et al.* Isolation and pharmacological action of epistephanine, an alkaloid of *Stephania hernandifolia.* Planta Med 1979; 35: 167-173.
28. Patra A, Mandal TK, Mukhopadhyay PK, *et al.* (+)-3′,4′-Dihydrostephasubine, a bisbenzylisoquinoline alkaloid from *Stephania hernandifolia. Phytochemistry* 1968; 27: 653-655.
29. Fadeeva I, Fesenko DA, Il'inskaya TN, *et al.* Alkaloids of *Stephania hernandifolia* VIII. Methylhernandine. *Chem Nat Comp* 1971; 7: 432-434.
30. Bose AN, Moza BK, Chaudhuri PK. Investigations on *Stephania hernandifolia* (Willd.) Walp. - V: Antispasmodic action of its constituents. *Ind J Pharm* 1968; 30: 179.

31. Mosihuzzaman M, Nahar N, Ali L, *et al.* Hypoglycemic effects of three plants from eastern Himalayan belt. *Diab Res* 1994; 26: 127-138.
32. Ghosh D, Jana D, Debnath JM. Effects of leaf extract of *Stephania hernandifolia* on testicular gametogenesis and androgenesis in albino rats: a dose-dependent response study. *Contraception* 2002; 65: 379-384.
33. Jana D, Maiti R, Ghosh D. Effect of *Stephania hernandifolia* leaf extract on testicular activity in rats. *Asian J Androl* 2003; 5: 125-129.

Chapter 15

Phytopharmacology of Indian Nootropic *Convolvulus plauricaulis* L.

15.1 INTRODUCTION

Convolvulus is a genus of about 250 species of flowering plants in the bindweed family Convolvulaceae, with a cosmopolitan distribution.[1] *C. pluricaulis* occurs in temperate and subtropical regions. *C. pluricaulis* is a perennial herb. The branches spread on the ground and can be more than 80 cm long. The leaves are elliptic in shape and flowers are blue in colors. The fruit is a nut; oblong, trigonous, stramineous and stiptate.[2,3]

An active alkaloid, sankhpuspine, has been isolated from *C. pluricaulis*. Volatile oil has been obtained by steam distillation of the fresh plant.[4,5] In Ayurveda, *C. pluricaulis* is used as medhya rasayana (nootropic or memory booster).[6] The syrup of the entire plant is a popular remedy for boosting childhood memory in India.

Therapeutically *C. pluricaulis* is considered to be tonic, alterative and febrifuge. It is used in treatment of fever, nervous debility and loss of memory. The whole plant is used medicinally in the form of decoction with cumin and milk in syphilis and scrofula.[7]

15.2 MATERIALS AND METHODS

C. pluricaulis is an important medicinal plant of Ayurvedic Materia Medica but not much has been written on its therapeutic potential. Databases including Annotated Bibliography of Indian Medicine, Ayush Health Portal, Pub Med, and Ayurvedic Pharmacopeia of India, Bibliography of Central Council of Research in Ayurveda and Siddha and indexed journals on

Ayurveda, phytotherapy and pharmacology were consulted for extracting data on *C. pluricaulis*.

15.3 RESULTS

15.3.1 Central Nervous System

15.3.1.1 Anti-anxiety activity

In the elevated plus maze, ethyl acetate fractions of *C. pluricaulis* at 100 mg/kg., PO showed an anxiolytic effect as evidenced by increase in the time spent in open arms and the number of open arm entries, compared to the control group. The ethyl acetate fractions of ethanol extract of the aerial parts at doses of 200 mg/kg PO significantly reduced the neuromuscular coordination indicative of the muscle relaxant activity at a higher dose in both the drugs.[8]

15.3.1.2 Anti-depressant activity

A. The alcoholic extract of *C. pluricaulis*, in a dose of 100 mg/kg body weight a barbiturate potentiation effect in albino rats; this effect was weaker than that of diazepam, but stronger than that of *Centella asiatica* (Linn.) Urban.[9]

B. The chloroform fraction of the total ethanolic extract of *C. pluricaulis* in doses of 50 and 100 mg/kg significantly reduced the immobility time in both forced swim test and tail suspension test. Its efficacy was found to be comparable to that of imipramine (15 mg/kg PO) and fluoxetine (20 mg/kg PO) administered for 10 successive days. The chloroform fraction reversed reserpine-induced extension of immobility period in forced swim test and tail suspension test. Prazosin, sulpiride, and p-chlorophenylalanine significantly attenuated the chloroform fraction-induced antidepressant-like effect in tail suspension test.[10]

15.3.1.3 Anti-epileptic activity

The alcoholic extract of *C. pluricaulis* stopped spontaneous motor activity and the fighting response, but did not affect the escape response; electrically induced convulsive seizures and tremorine-induced tremors were antagonized by the extract.[11]

15.3.1.4 CNS-depressant activity

The methanol extract of the whole plant of *C. pluricaulis* was found to produce alterations in the general behavior pattern, reduction in spontaneous motor activity, hypothermia, potentiation of pentobarbitone-sleeping time, reduction in the exploratory behavioral pattern, and suppression of aggressive behavior. The extract also showed an inhibitory effect on conditioned avoidance response and antagonism to amphetamine toxicity.[12]

15.3.1.5 Analgesic activity

The *C. pluricaulis* extract caused a reduction in the fighting behavior in mice but was devoid of analgesic activity although it potentiated morphine analgesia.[13]

15.3.1.6 Nootropic activity

Ethanolic extract of *C. pluricaulis* enhanced neuropeptide synthesis in the brains of laboratory animals. The brain protein was increased, indicating increased memory and acquisition efficiency.[12]

15.3.2 Heart

15.3.2.1 Cardiovascular activity

An ethanolic extract of the whole plant exerted a negative ionotropic action on amphibian and mammalian myocardium. It also exerted spasmolytic activity on smooth muscles of blood vessels.[14]

15.3.2.2 Hypolipidemic activity

The ethanolic extract of *C. pluricaulis* reduced total serum cholesterol, triglycerides, phospholipids and non-esterified fatty acids after 30 days of administration in hyperlipidemic rats. Moreover, high density lipids were significantly raised in animals.[15]

15.3.3 Miscellaneous Activity

15.3.3.1 Antifungal activity

The extract of leaves and flowers, probably due to some flavones was found to possess antifungal property.[16]

15.3.3.2 Immunomodulatory activity

For the study, Freund's adjuvant was used to induce inflammation in the right hind paw of the animals in the study, and subsequently, the crude extract of *C. pluricaulis* was administered intraperitonially. The results indicated that treatment with *C. pluricaulis* significantly reduced both inflammation and edema. The animals treated with *C. pluricaulis* also showed mild synovial hyperplasia and infiltration of mononuclear cells.[17]

15.3.3.3 Thyrotropic activity

L-Thyroxine treatment for 30 days increased serum concentrations of thyroxine (T4) and triodothyronine (T3). The activity of hepatic 5′-monodeiodinase (5′-DI) and glucose-6-phosphatase (G-6-Pase) was also enhanced. On the other hand, administration of the *C. pluricaulis* extract either alone or with

L-T4, decreased serum T3 concentration and the activity of hepatic 5′-DI and G-6-phase, without marked alteration in hepatic lipid peroxidation, indicating the possible regulation of hyperthyroidism by the *C. pluricaulis* extract.[18]

15.3.4 Drug Interactions

During the course of routine plasma drug level monitoring, an unexpected loss of seizure control and reduction in plasma phenytoin levels was noticed in two patients who were also taking *C. pluricaulis*. Single dose *C. pluricaulis* and phenytoin (oral/IP) co administration did not have any effect on plasma phenytoin levels but decreased the anti-epileptic activity of phenytoin significantly. On multiple-dose co administration, *C. pluricaulis* reduced not only the anti-epileptic activity of phenytoin but also lowered plasma phenytoin levels. *C. pluricaulis* itself showed significant anti-epileptic activity compared to a placebo and is worth further investigation.[19]

15.3.5 Conclusions

Animal studies have demonstrated that the extract of *C. pluricaulis* possesses a range of pharmacological effects with regard to the central nervous system. Keeping in mind the preliminary data accumulated from various animal studies, clinical studies are warranted to unearth the therapeutic potential of the *C. pluricaulis*.

REFERENCES

1. Feinbrun-Dothan N. *Flora Palaestina*, Jerusalem Academic Press.Vol., 3, 1978; pp. 162-163.
2. Madhavan V. Yoganarasimhan N, Gurudeva MR. Pharmacognostical studies on Sankhapushpi (*Convolvulus microphyllus* Sieb. ex Spreng and *Evolvulus alsinoides* (L.) *Indian J Trad Knowl* 2008; 7: 529-541.
3. Karandikar GK, Satakapan S. Shankhapushpi-{A} pharmacognostic study--{II}. *Convolvulus microphyllus* Sieb. *Indian J Pharmacol* 1959: 204-207.
4. Basu NK, Dandiya PC. Chemical investigation of *Convolvulus pluricaulis* Choisy. *J American Pharm Assoc* 2006; 37: 27-28.
5. Bisht NPS, Singh R. Chemical Studies of *Convolvulus microphyllus* Sieb. *Planta Medica* 1978; 34: 222-223.
6. Dandiya PC, Chopra YM. CNS-active drugs from plants indigenous to India. *J Ethnopharmacol* 1970; 2: 67-90.
7. Singh AP. *Dravyaguna Vijnana.* Chaukhambha Orientalia, New Delhi, 2005; pp. 123.

8. Nahata A, Patil UK, Dixit VK. Anxiolytic activity of *Evolvulus alsinoides* and *Convolvulus pluricaulis* in rodents. *Pharma Biol* 2009; 47: 444-1451.
9. Shukla SP. A comparative study on the barbiturate hypnosis potentiation effect of medhya rasayana drugs-shankhapushpi (*Convolvulus pluricaulis*) *Bull Med Ethnobot Res*. 1980; 1: 554.
10. Dhingra D, Valecha R. Evaluation of the antidepressant-like activity of *Convolvulus pluricaulis* choisy in the mouse forced swim and tail suspension tests. *Int Med J Exp Clin Res* 2007; 13: 155-161.
11. Mulchandani MA, Barve A, Gokhale PC, *et al*. Mechanism of anti-epileptic effect of *Convolvulus pluricaulis*: effect on veratrine-induced amino acid release. *Indian J Pharmacol*. 1995; 27: 48 (abstract no. 19).
12. Pawar SA, Dhuley JN, Naik SR. Neuropharmacology of an extract derived from *Convolvulus microphyllus*. *Pharma Biol* 2001; 39: 253-258.
13. Sharma VN, Barar FSK, Khanna NK, *et al*. Some pharmacological actions of *Convolvulus pluricaulis* Choisy. An Indian indigenous herb, part II. *Indian J Med Res* 1965; 53: 871-876.
14. Chaturvedi GN, Sharma RK, Sen SP. Hypotensive effect of certain indigenous drugs with special reference to shankhapuspi (*C. pluricaulis*) in anaesthetized dogs. *J Res Indian Med* 1964; 2: 57-67.
15. Sinha SN, Dixit VP, Madnawat AVS, *et al*. Hypolidemic effects of the ethanolic extract of *C. pluricaulis*. *Indian Med* 1989; 1: 1-2.
16. Gupta RC, Mudgal V. Antifungal effect of *Convolvulus pluricaulis* (shankhpushpi). *J Res Indian Med* 1974; 9: 67-68.
17. Ganju L, Karan D, Chanda S, *et al*. Immunomodulatory effects of agents of plant origin. *Biomed Pharmacother* 2003; 57: 296-300.
18. Panda S, Kar A. Inhibition of T3 production in levothyroxine-treated female mice by the root extract of *Convolvulus pluricaulis*. *Horm Metabol Res* 2001; 1: 16.
19. Dandekar UP, Chandra RS, Dalvi SS, *et al*. Analysis of a clinically important interaction between phenytoin and Shankhapushpi, an Ayurvedic preparation. *J Ethnopharmacol* 1992; 35: 285-288.

Chapter 16

Viscum album Linn.—Valuable Anticancer Herbal Drug

16.1 INTRODUCTION

Viscum album Linn. (Loranthaceae) is a species of mistletoe, also known as European Mistletoe or Common Mistletoe to distinguish it from other related species. It is native to Europe and western and southern Asia.[1]

16.2 BOTANY

It is a hemi-parasitic shrub, which grows on the stems of other trees. It has stems 30-100 cm long with dichotomous branching. The leaves are in opposite pairs, strap-shaped, entire, leathery textured 2-8 cm long and 0.8-2.5 cm broad, yellowish-green in color. This species is diecious and the flowers are inconspicuous, yellowish-green, 2-3 mm in diameter. The fruit is a white or yellow berry containing one seed embedded in very sticky, glutinous fruit pulp.[2]

16.3 PHYTOCHEMISTRY

The toxic lectin Viscumin has been isolated from *Viscum album*.[3] Viscumin is a cytotoxic protein that binds to galactose residues of the cell surface glycoproteins and may be internalized by endocytosis. In addition, it contains triterpenoid saponin, choline, proteins, resin, mucilage, histamine, traces of an alkaloid. Recently, acyclic monoterpene glycoside has been isolated from the plant growing in Turkey.[4]

16.4 REVIEW OF MEDICINAL PROPERTIES

Mistletoe not only has an interesting mythology, but also is interesting from a medicinal point of view. Though the Druids probably overrated the herb, deeming it useful for any kind of ailment, later herbalists still valued it highly for a variety of different ailments. Most notably it is recommended as a remedy for epilepsy, especially childhood epilepsy. This treatment reflects a homeopathic approach, as large doses of the herb and in particular of the berries actually causes fits and convulsions. It was used specifically for this ailment and also as a nervine to treat hysteria, delirium, convulsions, neuralgia as well as urinary disorders and heart complaints especially when these are related to a nervous condition.

Mistletoe is also known as a cardioactive agent that improves the pulse, regulates the heart rate and simultaneously dilates the blood vessels, thus lowering blood pressure. It reduces headaches and dizziness caused by high blood pressure. However, from the available literature, it is not entirely clear in which form mistletoe should be administered for this effect. Some sources maintain that the cardioactive principle is only effective if applied by injection while other sources recommend standard teas, tinctures, and extracts.

One source also states that the active principles would be destroyed by heat and thus should only be prepared by cold infusion. Differing opinions regarding the preparation methods are certainly confusing. Recently another interesting property of mistletoe has become a matter of scientific interest. Since ancient times, mistletoe has been used to treat tumors. Modern animal studies have demonstrated antioxidant effects.[5]

16.5 MECHANISM OF ANTICANCER ACTION

The biological efficacy of mistletoe lectin can be regarded basically as directly cytostatic as well as having an immunomodulatory effect. In cultures of human peripheral mononuclear cells (PBMCs), VAA-I can stimulate cytokine production as well as programmed cell death (apoptosis) in approximately the same concentration as *in vivo*.

A ß-galactoside-specific lectin from a proprietary mistletoe extract, induced increased secretion of tumor necrosis factor *α*, interleukin 1, and interleukin 6 in cultures of human peripheral blood mononuclear cells. The enhancement of secretion, determined independently by bioassays and enzyme-linked immunosorbent assay-based quantitation, was caused by selective protein-carbohydrate interaction, as revealed by the strict dependence on the presence of the carbohydrate-binding subunit of the lectin and the reduction of the effect of the lectin in the presence of the specific lectin-binding sugar as well as anti-lectin antibodies. Increased cytokine levels in the serum of patients after injection of optimal lectin doses corroborated the *in vitro* results.[5]

The *in vitro* effects of therapeutically administered mistletoe extracts (ABNOBAviscum) and pure mistletoe lectins (mainly mistletoe lectin I) on a variety of human and murine tumor cell lines were investigated. Mistletoe extracts and purified mistletoe lectins inhibited *in vitro* growth of all tumor cell lines tested including B cell hybridomas, P815, EL-4, Ke37, MOLT-4 and U937. The mechanism of growth arrest was shown to be due to the induction of programmed cell death (apoptosis). Thus, fragmentation of genomic DNA into oligonucleosomal bands of approximately 200 base pairs in length was observed within 20 hr when tumor cells were incubated with mistletoe extracts or lectins. The data points to a rational basis for the direct cytotoxic effects of mistletoe extracts and lectins apart from the postulated immunostimulatory properties of these agents.[6]

The present work examined the cytotoxic effects of *Viscum album* extracts produced from mistletoes grown on different host trees and of purified toxic proteins from VAL, such as the D-galactose-specific lectin I (ML I), the N-acetyl-D-galactosamine-specific ML II and ML III, and crude viscotoxins towards cultured human lymphocytes. Using the purified proteins, it became obvious that the cell killing was mediated by the induction of apoptosis, as measured by the appearance of a hypodiploid DNA peak using flow cytometry. ML III was the most effective to induce apoptosis, followed by ML II and ML I, while the viscotoxins and oligosaccharides from VAL did not. The findings suggest that there might be at least two different ways of killing the cell that is operative in VAL-mediated cytotoxicity:

(a) Typical apoptotic cell death with the appearance of hypo-diploid nuclei.

(b) A direct or indirect killing by damaging the cell membrane with subsequent influx of Ca(2+) and of the DNA intercalating dye propidium iodide and cell shrinkage. These effects might not be exclusive, as they probably occur simultaneously.[7]

Mistletoe lectin I from *Viscum album* inhibited cell growth and induces apoptosis (programmed cell death) in several cell types. Because increases in cytosolic Ca(2+) concentration ([Ca2+]i) constitute a signal for the induction of apoptosis, a study investigated the effects of Mistletoe lectin I on basal [Ca(2+)]i, receptor-mediated rises in [Ca2+]i and cell viability, using human U-937 promonocytes as model system.

Treatment of U-937 cells with ML I (30-100 ng/ml) significantly increased basal [Ca2+]i. Mistletoe lectin I (10-30 ng/ml) enhanced histamine-induced rises in [Ca2+]i up to five-fold. The effect of histamine was inhibited by clemastine but not by famotidine, indicative of its mediation via H1-receptors. Mistletoe lectin I additionally enhanced the stimulatory effect of complement C5a on [Ca2+]i, whereas the effect of ATP was unaffected. Mistletoe lectin I up to 10 ng/ml did not affect cell viability and growth of U-937 cells. Mistletoe lectin II at 30 ng/ml moderately inhibited cell growth and reduced cell viability. At 100 ng/ml, Mistletoe lectin I was strongly cytotoxic. Data suggests that Ca(2+) plays a role as intracellular signal molecule in the induction of apoptosis and points to an accelerating role of H1- and C5a-receptors in the regulation of this process.[8]

Mistletoe lectin I (ML I) is a major active component in plant extracts of *Viscum album* that is increasingly used in adjuvant cancer therapy. ML I exerts potent immunomodulating and cytotoxic effects, although its mechanism of action is largely unknown. Researchers showed that treatment of leukemic T- and B-cell lines with ML I induced apoptosis, required the prior activation of proteases of the caspase family.

The involvement of caspases is demonstrated because

(a) A peptide caspase inhibitor almost completely prevented ML I-induced cell death and

(b) Proteolytic activation of caspase-8, caspase-9, and caspase-3 was observed.

Because caspase-8 has been implicated as a regulator of apoptosis mediated by death receptors, we further investigated a potential receptor involvement in ML I-induced effects. ML I triggered a receptor-independent mitochondria-controlled apoptotic pathway because it rapidly induced the release of cytochrome *c* into the cytosol. Because ML I was also observed to enhance the cytotoxic effect of chemotherapeutic drugs, these data may provide a molecular basis for clinical trials using MLs in anticancer therapy.[9, 10]

REFERENCES

1. Zuber D. Biological flora of Central Europe: *Viscum album* L. *Flora* 2004; 199: 181-203.
2. Blamey M, Grey-Wilson C. *The Illustrated Flora of Britain and Northern Europe.* Hodder & Stoughton; 1989.
3. Stirpe F, Sandvig K, Olsnes S, *et al.* Action of viscumin, a toxic lectin from mistletoe, on cells in culture. *J Biol Chem* 1982; 257: 13271-13277.
4. Deliorman D, Ergun CF. A new acyclic monoterpene glucoside from *Viscum album* ssp. *Album. Fitoterapia* 2001; 72: 101-105.
5. Onay-Ucar E, Karagoz A, Arda N. Antioxidant activity of *Viscum album. Fitoterapia* 2006; 77: 556-560.
6. Hajto T, Hostanska K, Frei K, *et al.* Increased secretion of tumor necrosis factor α, interleukin 1, and interleukin 6 by human mononuclear cells exposed to ß-galactoside-specific lectin from clinically applied misletoe extract *Cancer Res* 1990; 50: 3322-3326.
7. Janssen O, Scheffler A, Kabelitz D. *In vitro* effects of mistletoe extracts and mistletoe lectins. Cytotoxicity towards tumor cells due to the induction of programmed cell death (apoptosis) *Drug Res* 1993; 43: 1221-1227.
8. Büssing A, Suzart K, Bergmann J, *et al.* Induction of apoptosis in human lymphocytes treated with *Viscum album* L is mediated by mistletoe lectins *Cancer Lett* 1996; 99: 59-72.

9. Wenzel-Seifert K, Lentzen H, Seifert R. In U-937 peomonocytes, mistletoe lectin I increases basal [Ca2+]i, enhances histamin H1- and complement C5a-receptor-mediated rises in [Ca2+]i, and induces cell death *Naunyn-Schmiedeberg's Arch Pharmacol* 1997; 355: 190-197.
10. Bantel H, Engels I, Voelter W, *et al.* Mistletoe lectin activates caspase-8/ FLICE independently of death receptor signaling and enhances anticancer drug-induced apoptosis *Cancer Res* 1999; 59: 2083-2090.

Chapter 17

Anti-inflammatory Activity of Aqueous-Ethanolic Extract of Aerial Parts of *Artemisia vulgaris* Linn. in Spargue Dawley Rats

Singh AP and Duggal S*

17.1 INTRODUCTION

The genus *Artemisia* (family Asteraceae, tribe Anthemideae), comprises a variable number of species (from 200 to over 400, depending on the authors) found throughout the northern part of the world. *Artemisia vulgaris* L. (Asteraceae), commonly known as mugwort or common wormwood, is a perennial herb, widely distributed in different habitats, from 0 to 1800 m above sea level.[1]

In Ayurveda, *A. vulgaris* is source of snake-bite antidote drug "nagdaun".[2] In traditional herbal medicine, aerial parts of *A. vulgaris* are used as anthelmintic, antiseptic, antispasmodic, antidiabetic, antiepileptic and antidepressant.[3] It is widely used in the Philippines for its anti-inflammatory properties.[4]

A. vulgaris contains essential oils containing cineole and thujone.[5-7] Monoterpene and sesquiterpene lactones including vulgarin, pilostachyin and pilostachyin C have been reported.[8,9] The derivatives of quercetin and quercetagetin are the main flavonoids in acetone extract of leaf exudates from *A. vulgaris*.[10]

Two new eudesmane acids and a known eudesmane dialcohol and polyacetylenes have been isolated.[11-13] Dicaffeoylquinic acids, new sesqui-

Dept of Pharamacology, Rayat Bahara Institute of Pharmacy, Hoshiarpur (Punjab) India.

terpene, caryophyllene oxide, phytyl fatty acid esters, squalene, stigmasterol and sitosterol have been reported from the dichloromethane extract of the air-dried leaves of *A. vulgaris*.[14,15]

A previous work reported antimicrobial, antihypertensive, anti-spasmodic and bronchodilator, hepatoprotective, antidepressant, xanthine oxidase inhibitor and anti-oxidant activities of *A. vulgaris* extracts.[16-22] Erio-dictyol and luteolin have been reported as the chief esterogenic flavonoids among 20 isolated from the plant.[23]

A previous study reported the presence of anti-inflammatory constituent in the water extract fractions of leaves of *A. vulgaris*.[17] In the present study, an attempt was made to investigate the anti-inflammatory activities of the aqueous-ethanolic (40:60) extract of aerial part of *A. vulgaris* in experimental animals.

17.2 MATERIALS AND METHODS

17.2.1 Plant Materials

The fresh aerial parts of The *Artemisia vulgaris* were collected in the month of February from the botanical garden of Lovely Professional University, Phagwara, district Kapurthala of Punjab and were authenticated by Dr. R.K. Chaturvedi, Scientist, Central Council of Research in Ayurveda and Siddha, Patna. Herbarium of plant specimen was deposited at Regional Research Centre, CCRAS, Patna (voucher no. CCARS/RRC/2010/34).

17.2.2 Preparation of the Extract

Freshly collected aerial parts of the plant were washed with water and chopped with a stainless steel knife. The fresh aerial part was extracted with soxhlet apparatus using (40:60) aqueous: ethanol till exhaustion. The extract was evaporated under reduced pressure by a rotary vacuum evaporator until all the solvent was removed to give an extract sample. The combined extract sample was dried naturally in the shade and preserved in desiccators for further use.

17.2.3 Drug and Chemicals

Carrageenan was purchased from Hi Media Laboratories Pvt. Ltd. Mumbai, India; eggs were bought from the local market. All the other used solvents and chemicals used were of analytical grade.

17.2.4 Anti-inflammatory Activity

17.2.4.1 Carrageenan-induced rat paw edema

Healthy Spargue Dawley rats were obtained from the central animal house, Lovely Professional University after approval from Institutional Animal

Ethics Committee vides approval number: 954/ac/06/CPCSEA/09/12. Experimental animals were divided randomly into six groups with five animals in each group.

Group I-Carrageenan
Group II-Carrageenan + Indomethacin 20 mg/kg
Group III-Carrageenan + AEE 50 mg/kg
Group IV-Carrageenan + AEE 100 mg/kg
Group V-Carrageenan + AEE 200 mg/kg
Group VI-Carrageenan + AEE 100 mg/kg + Indomethacin 20 mg/kg

The aqueous: ethanol extract of AEE was evaluated for its anti-inflammatory activity.[24] Acute inflammation was produced by sub-plantar injection of 0.1 ml of 1 percent carrageenan in 0.3 percent CMC in the right hind paw of the rats, 30 min after the oral administration of the drug/extract. The paw volume was measured by using plethysmometer, at the intervals of 0, 30, 60, 120, 180 and 240 min after the carrageenan injection.

17.2.4.2 Fresh egg albumin-induced paw edema

Group I-egg albumin (0.1 ml, without dilution)
Group II-egg albumin + Indomethacin 20 mg/kg
Group III-egg albumin + AEE 50 mg/kg
Group IV-egg albumin + AEE 100 mg/kg
Group V-egg albumin + AEE 200 mg/kg
Group VI-egg albumin + AEE 100 mg/kg + Indomethacin 20 mg/kg

The paw edema was induced by sub-plantar injection of 0.1 ml of fresh egg albumin in the right hind paw of the rats, 30 min after the oral administration of the drug/extract. Anti-inflammatory activity was measured during carrageenan-induced inflammation and percentage inhibition of inflammation was calculated using the following formula.[25]

Percent inhibition = {(edema volume in control – edema volume in treated)/edema volume in control} × 100

17.3 RESULTS AND DISCUSSION

The result of carrageenan-induced rat paw edema is shown in Table 17.1. The AEE extract at doses of 100 and 200 mg/kg shows reduction in paw volume (33.77 and 21.0 percent respectively) at 240 min, whereas at 50 mg/kg maximum reduction were (4.26 percent) after 30 min. In case of 100 mg/kg + 20 mg/kg indomethacin shows maximum inhibition (48.74 percent) after 240 min.

Carrageenan is the phlogistic agent of choice for testing anti-inflammatory substances as it is not known to be antigenic and is devoid of apparent systemic effects. Moreover, the experimental model exhibits a high degree of reproducibility. Carrageenan-induced rat paw edema was markedly inhibited by pretreatment at different dose level i.e. 50, 100, 200 mg/kg with respect to control animals.

Table 17.1 Anti-inflammatory activity of *A. vulgaris* using carrageenan-induced inflammation on hind paw of rats

Groups	0 min	30 min	60 min	120 min	180 min	240 min
Group I	0.87±0.010	0.94±0.006	1.08±0.006	1.11±0.010	1.13±0.011	1.19±0.009
Group II	0.80±0.014 (8.05%)	0.79±0.013 (15.96%)	0.70±0.013 (35.19%)	0.67±0.011 (39.63%)	0.68±0.010 (39.82%)	0.67±0.013 (43.69%)
Group III	0.86±0.035 (1.15%)	0.90±0.030 (4.26%)	1.04±0.024 (3.70%)	1.10.98±0.027 (0.90%)	1.12±0.031 (0.88%)	1.19±0.028 (0%)
Group IV	0.68±0.038 (21.84%)	0.71±0.037 (24.47%)	0.74±0.352 (31.48%)	0.76±0.035 (31.53%)	0.79±0.039 (30.09%)	0.80±0.039 (32.77%)
Group V	0.78±0.013 (10.34%)	0.80±0.016 (14.89%)	0.83±0.013 (23.15%)	0.85±0.007 (23.42%)	0.86±0.010 (23.89%)	0.94±0.010 (21.01%)
Group VI	0.83±0.010 (4.60%)	0.74±0.006 (21.28%)	0.67±0.004 (37.96%)	0.64±0.004 (42.34%)	0.62±0.007 (45.13%)	0.61±0.007 (48.74%)

Values are expressed as mean ±SEM, n = 5, values given in the parenthesis are percentage change with respect to the control group of the respective time.

The effect on paw volume with time duration induced by fresh egg albumin-induced acute inflammation is shown in Table 17.2. The inflammation induced by fresh egg albumin in different groups was compared with control animals group, revealing that after 240 min at different doses moderate effect against acute inflammation was observed. At dose of 100 and 200 mg/kg the percentage protection against inflammation was same (22.22 percent). At dose of 100 mg/kg AEE + 20 mg/kg indomethacin the maximum effect was observed 24.14 (120 min), 28.57 (180 min) and 26.66 percent (240 min).

Table 17.2 Anti-inflammatory activities of *A. vulgaris* using fresh egg albumin-induced inflammation on hind paw of rats

Groups	0 min	30 min	60 min	120 min	180 min	240 min
Group I	0.76±0.793	0.80±0.079	0.84±0.077	0.87±0.068	0.91±0.064	0.90±0.029
Group II	0.61±0.011 (19.74%)	0.62±0.010 (22.5%)	0.62±0.010 (26.19%)	0.62±0.010 (28.74%)	0.62±0.012 (31.87%)	0.62±0.013 (31.11%)
Group III	0.68±0.055 (10.53%)	0.72±0.051 (10%)	0.76±0.045 (9.52%)	0.76±0.047 (12.64%)	0.77±0.046 (15.38%)	0.78±0.028 (13.33%)
Group IV	0.63±0.050 (17.11%)	0.66±0.034 (17.5%)	0.68±0.046 (19.05%)	0.68±0.049 (21.84%)	0.69±0.049 (24.18%)	0.70±0.039 (22.22%)
Group V	0.61±0.035 (19.74%)	0.64±0.034 (20%)	0.67±0.035 (20.24%)	0.67±0.031 (22.99)	0.70±0.031 (23.08%)	0.70±0.010 (22.22%)
Group VI	0.63±0.051 (17.11%)	0.64±0.054 (20%)	0.66±0.054 (21.43%)	0.66±0.052 (24.14%)	0.65±0.053 (28.57%)	0.66±0.077 (26.66%)

Values are expressed as mean ±SEM, n = 5, values given in the parenthesis are percentage change with respect to the control group of the respective time.

The early phase (1-2 hr) of carrageenan model is mainly mediated by histamine, serotonin and increased synthesis of prostaglandins in the damaged tissue surroundings. The latter phase is sustained by prostaglandin release and mediated by bradykinin, leukotrienes, polymorphonuclear cells and prostaglandins produced by tissue macrophages.[26,27]

17.4 CONCLUSION

The results of the present study confirmed the aqueous ethanolic extract of *A. vulgaris* aerial part, shows a protective effect against inflammation induced by carrageenan and fresh egg albumin-induced inflammation. Further studies are required to evaluate the chemical constituents responsible for activities.

REFERENCES

1. Valan-Vetschera K, Wollenweber E. Exudate flavonoid aglycones in the alpine species of Achillea sect. Ptarmica: Chemosystematics of *Artemisia moschata* and related species. *Biochem Syst Ecol* 2001; 29: 149-159.
2. Issar RK. A note on the botanical sources of a snake-bite antidote drug "nagdaun. *J Res Indian Med* 1975; 10: 171-172.
3. Duke JA, Godwin MJB, DuCellier J. Characterization of antioxidant activity of extract from *Artemisia vulgaris*. Handbook of medicinal herbs, 2nd ed. CRC press, Washington DC, 2002, pp, 56-58.
4. Tigno XT, Guzman FD, Flora AM. Phytochemical analysis and hemodynamic actions of *Artemisia vulgaris* L. *Clin Hemorheol Microcircul* 2000; 23: 167-175.
5. Nano GM, Bicchi C, Frattini C, *et al.* Composition of some oils from *Artemisia vulgaris. Planta Med* 1976; 30: 211.
6. Mishra MB, Singh SW, Misra RK, *et al.* Preliminary pharmacological screening of *Artemisia vulgaris* Linn. *Indian J Med Sci* 1968; 22: 141-143.
7. Michaelis K, Vostronsky O, Paulini H, *et al.* On the essential oil components from blossoms of *Artemisia vulgaris* L. *Verlag der Zeitschrift für Naturforschung* 1982; 37: 152.
8. Naf-Muller R, Pickenhagen W, Willhalm B. New Irregular Mono-terpenes in *Artemisia vulgaris. Helv Chim Acta* 1981; 64: 142-143.
9. Marco JA. Sesquiterpenes lactones from Artemisia species. *Phytochemistry* 1993; 32: 460-462.
10. Milenanikolova M. High performance liquid chromatography separation of surface flavonoid aglycone in *Artemisia annua* L. and *Artemisia vulgaris* L. *J Serbian Chem Soc* 2004; 69: 571-574.
11. Marco JA, Sanz JF, Hierro P. Two eudesmane acids from *Artemisia vulgaris. Phytochemistry* 1991; 30: 2403-2404.

12. Drake D, Lam J. Polyacetylenes of *Artemisia vulgaris. Phytochemistry* 1974; 13: 455-457.
13. Wallnofer B, Hofer O, Greger H. Polyacetylenes from *Artemisia vulgaris. Phytochemistry* 1989; 28: 2687.
14. Carnat A, Heitz A, Fraisse D, *et al.* Major dicaffeoylquinic acids from *Artemisia vulgaris. Fitoterapia* 2000; 71: 587-589.
15. Ragasa CY, DeJesus J, Apuada MJ, *et al.* A new sesquiterpene from *Artemisia vulgaris. J Nat Med* 2008; 2: 461-463.
16. Laxmi US, Rao JT. Antimicrobial Properties of the Essential oils of *Artemisia pallens* and *Artemisia vulgaris. J Cosm Perf Sci Technol* 1991; 72: 510-511.
17. Tigno X, Gumila E. *In vivo* micro vascular actions of *Artemisia vulgaris* L. in a model of ischemia-reperfusion injury in the rat intestinal mesentery. *Clin Hemorheol Microcircul* 2000; 23: 159-165.
18. Khan AH, Gilani AH. Antispasmodic and bronchodilator activities of *Artemisia vulgaris* are mediated through dual blockade of muscarinic receptors and calcium influx. *J Ethnopharmacol* 2009; 126: 480-486.
19. Gilani AH, Yaeesh, S, Jamal Q, *et al.* Hepatoprotective activity of aqueous-methanol extract of *Artemisia vulgaris. Phytother Res* 2005; 19: 170-172.
20. Lee SJ, Chung HY, Lee IK, *et al.* Phenolics with Inhibitory Activity on Mouse Brain Monoamine Oxidase (MAO) from Whole Parts of *Artemisia vulgaris* L (Mugwort). *Food Sci Biotechnol* 2005; 9: 179-182.
21. Thanh NM, Awale S, Tezuka Y, *et al.* Xanthine oxidase inhibitory activity of Vietnamese medicinal plants. *Biol Pharma Bull* 2004; 27: 1414-1421.
22. Temraz A, El-Tantawy WH. Characterization of antioxidant activity of extract from *Artemisia vulgaris.* Pakistan *J Pharma Sci* 2008; 21: 321-326.
23. Lee SJ, Chung HY, Maier CGA, *et al.* Estrogenic Flavonoids from *Artemisia vulgaris* L. *J Agricul Food Chem* 1998; 46: 3325-3329.
24. Winter CA, Risley EA, Nuss GW. Carrageenan-induced edema in hind paws of the rat as an assay for the anti-inflammatory drugs. *Proc Soc Exper Biol Med* 1962; 111: 544-547.
25. Chu D, Kovacs BA. Anti-inflammatory activity in Oak gall extracts. *Arch Int Pharmacodyn* 1977; 230: 166-176.
26. Vinegar R, Schreiber W, Hugo R. Biphasic development of carrageenan oedema in rats. *J Pharmacol Exp Ther* 1969; 66: 96-103.
27. Crunkhon P, Meacock S. Mediators of the inflammation induced in the rat paw oedema in rats by carrageenan. *British J Pharmacol* 1971; 42: 392-402.

Chapter 18

Pharmacological Update of Anthraquinones

18.1 INTRODUCTION

Traditional drugs have given an important lead in drug search, resulting in the discovery of novel molecules.[1] Anthracene glycosides are also known as anthracenosides. They are purgative in nature. On hydrolysis, they produce glycones like dianthrone, anthraquinone or anthrone. The sugars are arabinose, rhamnose or glucose. Anthraquinones (Fig. 18.1) are the active constituents and are responsible for the biological activity of the anthracene glycoside containing drugs. In addition to the use in treating constipation, they are also used for the treatment of skin disease like psoriasis and ringworm (Fig. 18.2).

Anthraquinones are one such class of organic compounds which serve as a basic building block for a number of naturally and synthetic derivatives, widely used as a colorants in food, drugs, cosmetics, hair dyes and textiles.[2,3] These derivatives, usually in the form of glycosides or rhamnosides, are the active components in a number of crude drugs having various pharmacological properties such as antimicrobial, antifungal, analgesic and antihypertensive, antimalarial, anti-oxidant) antileukemic and mutagenic functions.

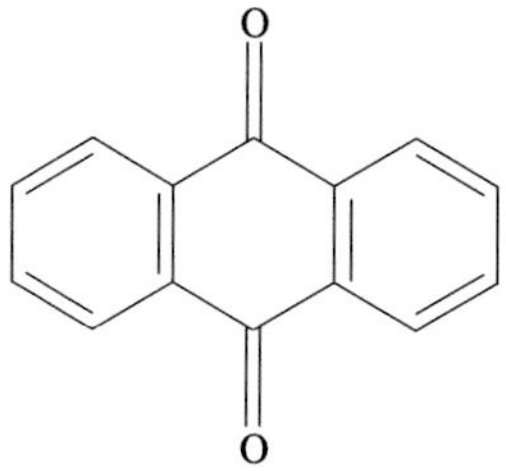

O

O

Figure 18.1 Structure of anthraquinone.

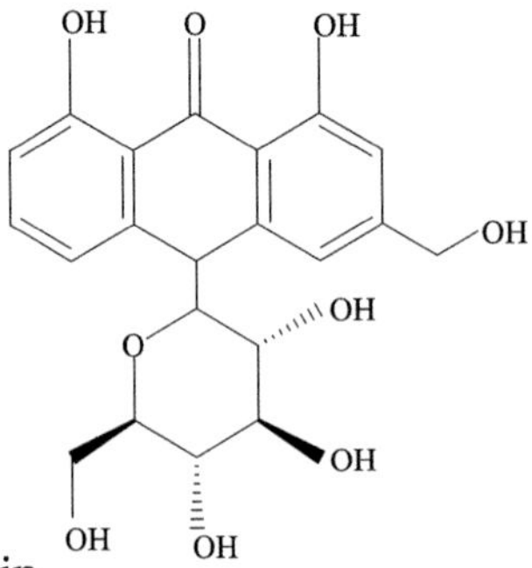

Aloin

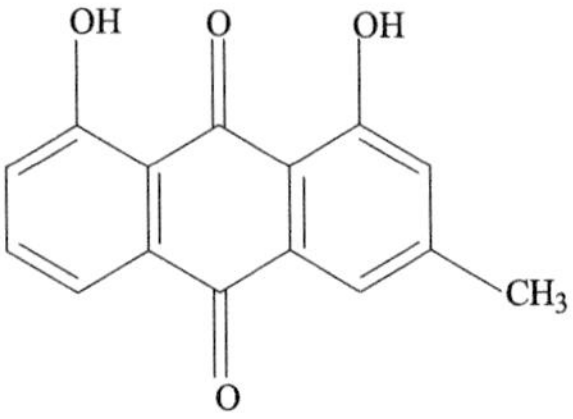

Chrysophanol

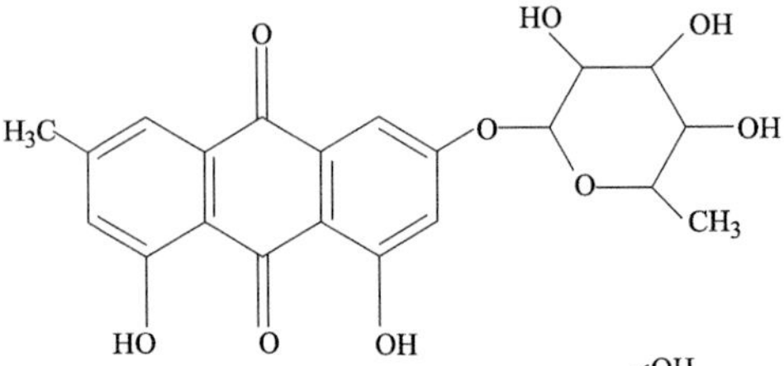

Frangulin A

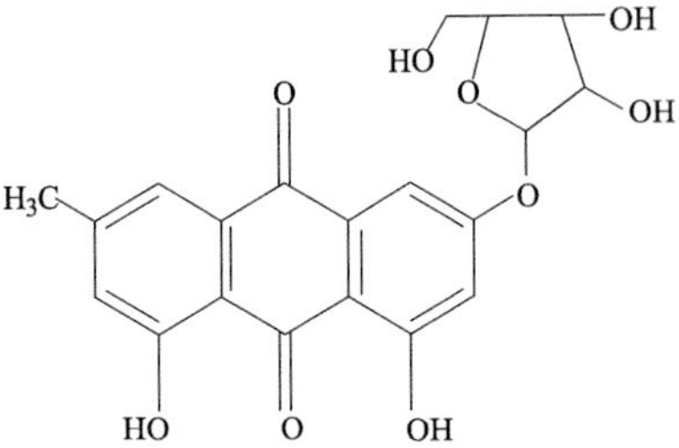

Frangulin B

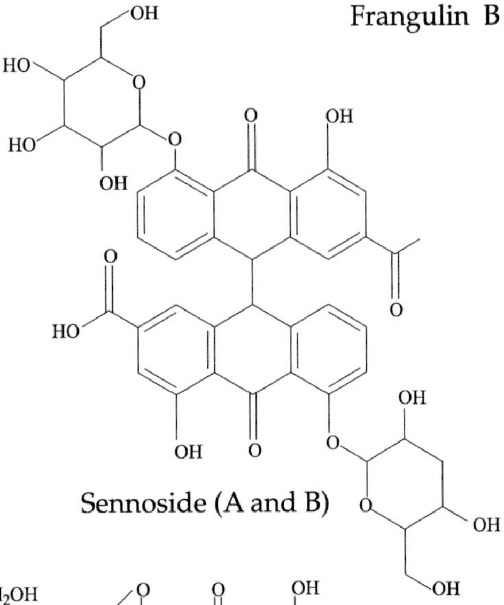

Sennoside (A and B)

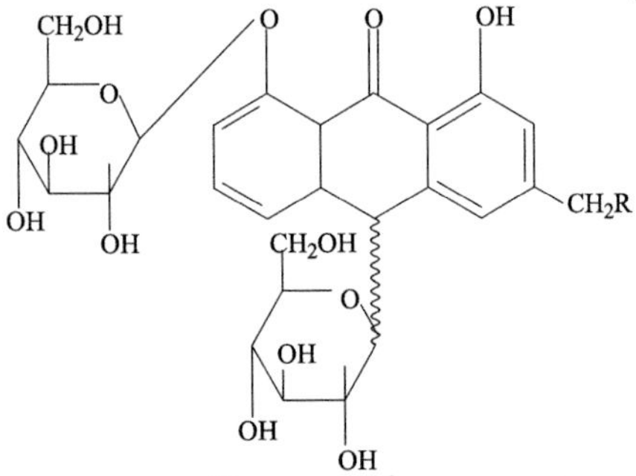

Cascarosides

Figure 18.2 Structure of common anthracene glycosides.

The presence of the sugar residue is a prerequisite for the pharmacological effects of anthraquinones, since the sugar moiety increases the solubility of the molecule and facilitates its transport to the site of action. They are hydrolyzed in the colon by the enzymes of the microflora.[4,5]

However, natural anthraquinones are distinguished by a large structural variety, a wide range of biological activity and low toxicity. Only a few medicinal plants have attracted the interest of scientists to investigate the remedy from anthraquinone. Hence, an attempt has been made to review some anthraquinone and its derivatives presents in medicinal plants used for the prevention and treatment of different disorders.

18.2 PHARMACOLOGICAL ACTIVITIES

18.2.1 Anticancer Activity

Anthraquinone monomers showed higher antitumor promoting activity than the bianthraquinones.[6] (10 anthraquinones rubiadin, rubiadin 1-methyl ether, lucidin, damnacanthol, 1,3-dihydroxy-2-methoxymethylanthraquinone, 3-hydroxy-1-methoxy-2-methoxymethylanthraquinone, nordamnacanthal, damnacanthal, sorandidiol and morindone were isolated from *Morinda citrifolia* Linn. root. 1,3-dihydroxy-2-methoxymethylanthraquinone, 3-hydroxy-1- methoxy-2-methoxymethylanthraquinone, nordamnacanthal, damnacanthal, sorandidiol and morindone exhibited remarkable inhibition against the activities of animal pols, and compound morindone found to be the strongest inhibitor investigated. The tendency of pol inhibition showed a positive correlation with the suppression of human colon cancer cell HCT116 growth and suggested that may be used as an anticancer functional food.[7]

Six anthraquinones, nordamnacanthal, alizarin-1-methyl ether, rubiadin, soranjidiol, lucidin-ω-methyl ether and morindone isolated from the cell suspension culture of *Morinda elliptica* were assayed for antitumor promoting and anti-oxidant activities at the concentration of 2.0 μg/ml. Morindone was found to be active as free radical scavenger with IC50 of 40.6 μg/ml.[8] *In vitro* assays for antitumor promoters were carried out for several derivatives chrysophanol, emodine, cassiamin C, cassiamin A, cassiamin B, 1,10,3,8,80-pentahydroxy-30,6-dimethyl (2,20-bianthracene)-9,90,10,100-tetrone, madagas-carin and 70-chloro-1,10,60,8,80-pentahydroxy-3, 30-dimethyl(2,20-bianthracene)-9,90,10,100-tetrone were isolated from *Cassia siamea*. Apart from cascaroside A, various derivatives isolated aloe-emodin, 1,10,8,80-tetrahydroxy-3,30-dihydroxymethyl(2,20-bianthracene)-9, 90,10,100-tetrone and 1,10,30,8,80-pentahydroxy-3-hydroxymethyl-60-methyl (2,20-bianthracene)-9,90,10,100-tetrone.[6]

Nordamnacanthal, damnacanthal and 1-hydroxy-2-hydroxymethyl-3-methoxyanthraquinone isolated from the roots and stems of *Prismatomeris fragrans* exhibited cytotoxicity to BC cell line while 1,3-dihydroxy-2-methyl-5,6-dimethoxy-anthraquinone, anthraquinones, nordamnacanthal,

damnacanthal, rubiadin, rubiadin-1-methyl ether and β-acetylolean-12-en-28-olic acid exhibited cytotoxicity to NCI-H187 cell line.[9]

Mollugin, 1-hydroxy-2-methylanthraquinone, 2-ethoxymethylanthraquinone, rubiadin, 1,3-dihydroxyanthraqunone, 7-hydroxy-2-methylanthraquinone, lucidin, 1-methoxymethyl anthraquinone and lucidin-3-O-primeveroside isolated from the roots of *Rubia tinctorum* showed mutagenicity with *Salmonella typhimurium* TA100 and/or TA98. The maximum antimutogenic activity was exhibited by 1,3-dihydroxyanthraquinones.[10]

Aloe emodin isolated from the seeds of *Rhamnus frangula* L., assayed for tumor-inhibitory activity. It was found that the compound showed significant antileukemic activity against the P-388 lymphocytic leukemia in mice.[11]

Aloesin, aloe-emodin and barbaloin extracted from *Aloe vera* leaves exhibited significant prolongation of the life span of tumor-transplanted animals. Also *A. vera* active constituents exhibited significant inhibition on Ehrlich ascite carcinoma cell number, when compared to the positive control group. Moreover, in trypan blue cell viability assay, active principles showed a significant concentration-dependent cytotoxicity against acute myeloid leukemia and acute lymphocytes leukemia cancerous cells. Furthermore, in MTT cell viability test, aloe-emodin was found to be active against two human colon cancer cell lines (i.e. DLD-1 and HT2), with IC50 values of 8.94 and 10.78 μM, respectively.[12]

In vivo mouse comet assay was performed on both isolated kidney and colon cells in order to demonstrate a possible organospecific genotoxicity after oral administration of Aloe emodin. It induced a clear genotoxic activity both in the *Salmonella typhimurium* strains TA1537 and TA98 and in the *in vitro* micronucleus assay in the absence as well as in the presence of metabolic activation.[13]

Alizarin (1,2-dihydroxyanthraquinone), isolated from *Rubia cordifolia* L. was evaluated for its antigenotoxic potential against a battery of mutagens viz. 4-nitro-o-phenylenediamine and 2-aminofluorene in Ames assay using TA98 tester strain of *Salmonella typhimurium*; hydrogen peroxide (and 4-nitroquinoline-1-oxide) in SOS chromotest using PQ37 strain of *Escherichia coli* and in Comet assay using human blood lymphocytes. The results showed that alizarin possessed a significant modulatory role against the genotoxicity of mutagens.[14]

Denbinobin, a 1,4-phenanthrenequinone, first isolated from the stems of *Dendrobium moniliforme* has been reported to exhibit antitumoral and anti-inflammatory activities.[15] Three new anthrone-anthraquinones, scutianthraquinones A, B and C, one new bisanthrone-anthraquinone, scutianthraquinone D, and the known anthraquinone, aloesaponarin isolated from an ethanolic extract of the bark of *Scutia myrtina* compounds were tested against the A2780 human ovarian cancer cell line for antiproliferative activities, and against the chloroquine-resistant *Plasmodium falciparum* strains Dd2 and FCM29 for antiplasmodial activities.

Compounds scutianthraquinones A, B and scutianthraquinone D showed weak antiproliferative activities against the A2780 ovarian cancer cell line, while all compounds except aloesaponarin I exhibited moderate antiplasmodial activities against *P. falciparum* Dd2 and compounds scutianthraquinones A, B and scutianthraquinone D exhibited moderate antiplasmodial activities against *P. falciparum* FCM29.[16]

18.2.2 Antimicrobial Effect

18.2.2.1 Antimalarial activity

10-(Chrysophanol-7′-yl)-10-(xi)-hydroxychrysopanol-9-anthrone and chryslandicin, isolated from the roots of *Kniphofia foliosa* showed a high inhibition on the growth of the malaria parasite, *P. falciparum* with ED50 values of 0.260 and 0.537 μg/ml, respectively. Bazouanthrone(3,5,8,9-tetrahydroxy-2,4,4-tri-(3,3-dimethylallyl)-6-methyl-1-(4H)anthracenone), an anthrone derivative, with known compounds, feruginin A, harunganin, harunganol A, harunganol B, friedelan-3-one and betulinic acid isolated from the root bark of *Harungana madagascariensis* were tested for antiplasmodial activity in culture against W2 strain of *Plasmodium falciparum*. All the compounds were found to be active against the Plasmodium parasites with bazouanthrone showing particular potency (IC50 = 1.80 μM).[17]

18.2.2.2 Antibacterial activity

2-(hydroxymethyl)anthraquinone, isolated from the dried inner bark of *Tabebuia impetiginosa* exhibited strong activity against *Helicobacter pylori* with the paper disc diffusion assay In the MIC bioassay, 2-(hydroxymethyl) anthraquinone (2 μg/ml), anthraquinone-2-carboxylic acid (8 μg/ml), and lapachol (4 μg/ml) were more active than metronidazole (32 μg/ml) but less effective than amoxicillin (0.063 μg/ml) and tetracycline (0.5 μg/ml).[18]

Febrifuquinone, a new vismione-anthraquinone coupled pigment and a new bianthrone named adamabianthrone, were isolated respectively, from the roots of *Psorospermum febrifugum* and from the bark of *Psorospermum adamauense* showed significant antimicrobial activities against a wide range of bacteria and fungi.[19]

Anthraquinone glycoside, rubiacordone A, isolated from the dried roots of *Rubia cordifolia* demonstared antimicrobial activities against Gram-positive and Gram-negative bacteria by the disc diffusion method.[20]

Zenkequinone B isolated from the stem bark of *Stereospermum zenkeri* exhibited a significant antimicrobial activity (MIC 9.50 μg/mL) against gram-negative *Pseudomonas aeruginosa*.[17] Nordamnacanthal, damnacanthal and morindone isolated from the roots of *Morinda elliptica* were found to have a strong antimicrobial activity.[21]

1,8-dihydroxy-2-methyl-3,7-dimethoxyanthraquinone, lucidin-3-O-β-primeveroside, 1,3-dihydroxy-2-methylanthraquinone, lucidin-ω-ethyl ether, lucidin-ω-butyl ether, damnacanthol identified from *Morinda angustifolia* demonstrated significant antimicrobial activity with the disc diffusion assay.[22]

Emodin isolated from the ethyl acetate extract of the leaves of *Cassia nigricans* Vahl demonstrated significant antimicrobial activity. The LC50* of the emodin was found to be 42.77 μg/ml.[23]

Anthroquinone glycosides isolated from the methanolic extract of *Thespesia populnea* flowers showed antibacterial activity in the lowest tested concentration of 62.5 μg/ml and 125 μg/ml and were found to be active in a dose dependent manner.[24]

Newbouldiaquinone, naphthoquinone-anthraquinone ether coupled pigment from *Newbouldia laevis* demonstrated antimicrobial activity comparable with standard antibiotics.[25]

18.2.2.3 Anti-HIV activity

Various anthraquinones substituted with OH, NH_2, SO_3Na, halogens, aromatic, amines, etc., were evaluated against HIV-1. Among them 1,2,5, 8-tetrahydroxyanthraquinone (EC50 = 3.29 μM), 1,2,4-trihydroxyanthraquinone (EC50 = 3.44 μM), 3,4-didhydroxy-9,10-dioxo-2-anthracene-sulfonic acid (EC50 = 3.48 μM) poly-phenolic and poly-sulfate substituted anthraquinones were found to be most effective. Amino or halogen substituted anthraquinones exhibit low activity. Hypericin, an anthraquinone dimer exhibited a potent anti-HIV-1 activity (EC50** = 0.44 μM).[26]

18.2.2.4 Antifungal activity

Rhein, physcion, aloe-emodin and chrysophanol isolated from *Rheum emodi* rhizomes exhibited antifungal activity against *Candida albicans, Cryptococcus neoformans, Trichophyton mentagrophytes* and *Aspergillus fumigatus* (MIC25-250 mg/ml).These pure compounds were compared with crude MeOH extract and anthraquinone derivatives were found to be more active.[27]

18.2.2.5 Antiviral activity

Chrysophanic acid isolated from *Dianella longifolia*, was found to inhibit the replication of poliovirus types II, III *in vitro* at a 50 percent effective concentration of 0.21 and 0.02 mg/ml respectively. Chrysophanic acid was compared with other isolated anthraquinones derivatives rhein, 1,8-dihydroxyanthraquinone, emodin and aloe-emodin and was found to be most effective against poliovirus type III.[28]

Anthraquinone, chrysophanol 8-O-D-glucoside isolated from *Rheum palmatum* exhibited significant activity against HBV with an IC50 value of 36.98 ± 2.28 μg/ml. Anthraquinone chrysophanol 8-O-D-glucosides was found to be the most active compound and suggested promising compound for the development of new anti-HBV drugs in the treatment of HBV infection.[29]

---

* LC50 = Lethal concentration

**EC50 = The half maximal effective concentration

18.2.3 Antihypertensive Activity (ACE Inhibitor Activity)

Glucoaurantioobtusin, anthraquinone glycoside, isolated from *Cassia tora* demonstrated significant ACE inhibitor activity with an IC50 value of 30.24 ± 0.20 μM.[30]

18.2.4 Anti-arthritic and Anti-inflammatory Activity

Anthraquinone had the most preventive anti-arthritic activity recorded as compared with anthracene, anthranilic acid and cinnamic acid isolated from *Aloe vera* gel (Davis *et al.*, 1986). Anthraquinones from *Rubia cordifolia* exhibited anti-inflammatory activity, anti-exudative effect and antiproliferative action during the rapid development of a model edema.[31]

18.2.5 Anti-oxidant Activity

Rubidianin, isolated from alcoholic extract of *Rubia cordifolia* demonstrated significant anti-oxidant activity. It prevented lipid peroxidation induced by ferrous sulfate and t-butylhydroperoxide in a dose-dependent manner. The anti-oxidant activity of rubidianin was found to be more when compared with mannitol, vitamin E and p-benzoquinone.[32]

18.2.6 Anti-osteoporotic Activity

Rubiadin-1-methyl ether and 2-hydroxy-1-methoxy-anthraquinone, isolated from ethanolic extract of the roots of *Morinda officinalis* promoted osteoblast proliferation, while 1,2-dihydroxy-3-methylanthraquinone and 1,3,8-trihydroxy-2-methoxy-anthraquinone increased osteoblast ALP activity. All of the isolated compounds inhibited osteoclast TRAP activity and bone resorption. But physicion and 2-hydroxymethyl-3-hydroxyanthraquinone exhibited maximum inhibitory effects on osteoclastic bone resorption.[33]

18.2.7 Cognition-enhancing Activity

Glucoobtusifolin (1, 2, and 4 mg/kg, PO) and obtusifolin (0.25, 0.5, 1, and 2 mg/kg, PO) isolated from the seeds of *Cassia obtusifolia* L., significantly reversed scopolamine-induced cognitive impairments in the passive avoidance test. Both compounds improved escape latencies, swimming times in the target quadrant, and crossing numbers in the zone where the platform previously existed in the Morris water maze test. Furthermore, the compounds demonstrated inhibitory activity against acetylcholinesterase *in vitro*, glucoobtusifolin and obtusifolin (IC50 = 37.2 and 18.5 μM, respectively) and *ex vivo*.[34]

18.2.8 Hepatoprotective Activity

Four novel anthraquinone-benzisochromanquinone dimers, named floribundiquinones A, B, C, and D, along with six known anthraquinones, 10-(chrysophanol-7′-yl)-10-hydroxychrysophanol-9-anthrane, physcion, chry-sophanol, 1,5,8-trihydroxy-3-methyl-anthraquinone, aloe-emodin and xanth-orin were isolated from the roots of *Berchemia floribunda*. Hepatoprotective activities were evaluated against D-galactosamine-induced toxicity in WB-F344 rat hepatic epithelial stem-like cells.[35]

18.2.9 Laxative Activity

Herbs containing anthraquinone derivatives (rhubarb, senna, frangula, cascara, aloe) are used as laxatives. Emodin, iso-emodin, aloe-emodin, and chrysophanol obtained from extracts of *Cascara sargada*, when administered separately, has little purgative effect. Yet the three given in admixture produces a good purgative action. The leaf and flower of *Cassia podocarpa* having anthraquinone glycosides, free aglycone showed laxative activity.[36]

18.2.10 Miscellaneous Activities

In addition, laxative and diuretic activities are also reported for anthraquinones.[37,38] Plants containing 1-hydroxyanthraquinone have been widely used for pharmaceutical purposes such as treatment of kidney and bladder stones, as a laxative mixture, and as a mild sedative.[3,39]

REFERENCES

1. Mazumder PM, Percha V, Farswan M, *et al*. Cassia: A wonder gift to medicinal sciences. *Int J Comm Pharm* 2008; 1: 16-38.
2. Brown JP, Brown RJ. Mutagenesis by 9,10-anthraquinone derivatives and related compounds in *Salmonella typhimurium*. *Mutat Res* 1976; 40: 203-224.
3. Mori H, Yoshimi N, Iwata H, *et al*. Carcinogenicity of naturally occurring 1-hydroxyanthraquinone in rats: Induction of large bowel, liver and stomach neoplasms. *Carcinogenesis* 1990; 11: 799-802.
4. Robber JE, Speedie MK, Tyler VE. *Pharmacognosy and Pharmacobiotecnology*. William and Wilkin, 1996; 51.
5. Samuelsson G. *Drugs of Natural Origin*, 4th revised ed. Apotekarsocieteten. 1999; 190-194.
6. Koyama J, Morita I, Tagahara K, *et al*. Inhibitory effects of anthraquinones and bianthraquinones on Epstein-Barr virus activation. *Cancer Lett* 2001; 170: 15-18.

7. Kamiya K, Hamabe W, Tokuyam S, *et al.* Inhibitory effect of anthraquinones isolated from the Noni (*Morinda citrifolia*) root on animal A-, B- and Y-families of DNA polymerases and human cancer cell proliferation. *Food Chem* 2010; 118: 725-730.
8. Jasril Lajis NH, Mooi LY, Abdullah MA, *et al.* Antitumor promoting and antioxidant activities of anthraquinones isolated from the cell suspension culture of *Morinda elliptica. Asia Pac J Mol Biol Biotechnol* 2003; 11: 3-7.
9. Kanokmedhakul K, Kanokmedhakul S, Phatchana R. Biological activity of Anthraquinones and Triterpenoids from *Prismatomeris fragrans. J Ethnopharmacol* 2005; 100: 284-288.
10. Kawasaki Y, Goda Y, Yoshihira K. The mutagenic constituents of *Rubia tinctorum. Chem Pharm Bull (Tokyo)* 1992; 40: 1504-1509.
11. Kupchan SM, Karim A. Tumor inhibitors. Aloe emodin: antileukemic principle isolated from *Rhamnus frangula* L. *Lloydia* 1976; 39: 223-224.
12. El-Shemy HA, Aboul-Soud MA, Nassr-Allah AA, *et al.* Antitumor properties and modulation of antioxidant enzymes' activity by *Aloe vera* leaf active principles isolated via supercritical carbon dioxide extraction. *Curr Med Chem* 2010; 17: 129-138.
13. Nesslany F, Simar-Meintières S, Ficheux H, *et al.* Aloe-emodin-induced DNA fragmentation in the mouse *in vivo* comet assay. *Mutat Res* 2009; 678: 13-19.
14. Kaur P, Chandel M, Kumar S, *et al.* Modulatory role of alizarin from *Rubia cordifolia* L. against genotoxicity of mutagens. *Food Chem Toxicol* 2010; 48: 320-325.
15. Sánchez-Duffhues G, Calzado MA, de Vinuesa AG, *et al.* Denbinobin inhibits nuclear factor-kappa B and induces apoptosis via reactive oxygen species generation in human leukemic cells. *Biochem Pharmacol* 2009; 77: 1401-1409.
16. Hou Y, Cao S, Brodie PJ, *et al.* Antiproliferative and antimalarial anthraquinones of *Scutia myrtina* from the Madagascar forest. *Bioorg Med Chem* 2009; 17: 2871-2876.
17. Lenta BN, Weniger B, Antheaume C, *et al.* Anthraquinones from the stem bark of *Stereospermum zenkeri* with antimicrobial activity. *Phytochemistry* 2007; 68: 1595-1599.
18. Park BS, Lee HK, Lee SE, *et al.* Antibacterial activity of *Tabebuia impetiginosa* Martius ex DC (Taheebo) against *Helicobacter pylori. J Ethnopharmacol* 2006; 105: 255-262.
19. Tsaffack M, Nguemeving JR, Kuete V, *et al.* Two new antimicrobial dimeric compounds: febrifuquinone, a vismione-anthraquinone coupled pigment and adamabianthrone, from two Psorospermum species. *Chem Pharm Bull (Tokyo)* 2009; 57: 1113-1118.
20. Xiang Li, Liu Z, Chen Y, *et al.* Rubiacordone A: A new anthraquinone glycoside from the roots of *Rubia cordifolia. Molecules* 2009; 14: 566-572.

21. Ali AM, Ismail NH, Mackeen MM, *et al.* Antiviral, cyotoxic and antimicrobial activities of anthraquinones isolated from the roots of *Morinda elliptica. Pharm Biol* 2000; 38: 298-301.
22. Xiang W, Song QS, Zhang HJ, *et al.* Antimicrobial anthraquinones from *Morinda angustifolia. Fitoterapia* 2008; 79: 501-504.
23. Ayo RG, Amupitan JO, Zhao Y. Cytotoxicity and antimicrobial studies of 1,6,8-trihydroxy-3-methyl-anthraquinone (emodin) isolated from the leaves of *Cassia nigricans* Vahl. *Afr J Biotechnol* 2000; 6: 1276-1279.
24. Saravanakumar A, Venkateshwaran K, Vanitha J, *et al.* Evaluation of antibacterial activity, phenol and flavonoid contents of *Thespesia populnea* flower extracts. *Pak J Pharm Sci* 2009; 22: 282-286.
25. Eyong KO, Folefoc GN, Kuete V, *et al.* Newbouldiaquinone A: A naphth-oquinone-anthraquinoneether coupled pigment, as a potential antimicrobial and antimalarial agent from *Newbouldia laevis. Phytochemistry* 2010; 67: 605-609.
26. Chu CK, Schinazit RF, Nasr M. Anti-HIV-1 activities of anthraquinone derivatives *in vitro. Int Conf AIDS* 1989; 5: 561.
27. Agarwal SK, Singh SS, Verma S, *et al.* Antifungal activity of anthraquinone derivatives from *Rheum emodi. J Ethnopharmacol* 2000; 72: 43-46.
28. Semple SJ, Pyke SM, Reynolds GD, *et al.* Flower *in vitro* antiviral activity of the anthraquinone chrysophanic acid against poliovirus. *Antiviral Res* 2001; 49: 169-178.
29. Lia Z, Lic LJ, Suna Y, *et al.* Identification of natural compounds with Anti-Hepatitis B virus activity from *Rheum palmatum* L. ethanol extract. *Chemotherapy* 2007; 53: 320-326.
30. Hyun SK, Lee H, Kang SS, *et al.* Inhibitory activities of *Cassia tora* and its anthraquinone constituents on angiotensin-converting enzyme. *Phytother Res* 2008; 23: 178-184.
31. Davis RH, Patrick S, Agnew BS, *et al.* Antiarthritic activity of anthraquinones found in *Aloe vera* for podiatric medicine. *Journal of the American Podiatric Medical Assoc* 1986; 76: 61-66.
32. Cai Y, Sun M, Xing J, *et al.* Antioxidant phenolic constituents in roots of *Rheum officeinale* and *Rubia cordifolia*: structure-radical scavenging activeity relationships. *J Agric Food Chem* 2004; 52: 7884-7890.
33. Wu YB, Zheng CJ, Qin LP, *et al.* Antiosteoporotic activity of anthraquinones from *Morinda officinalis* on osteoblasts and osteoclasts. *Molecules* 2008; 14: 573-583.
34. Kim DH, Hyun SK, Yoon BH, *et al.* Gluco-obtusifolin and its aglycon, obtusifolin, attenuate scopolamine-induced memory impairment. *J Pharmacol Sci* 2009; 111: 110-116.
35. Wei X, Jiang JS, Feng ZM, *et al.* Anthraquinone-benzisochromanquinone dimers from the roots of *Berchemia floribunda. Chem Pharm Bull (Tokyo)* 2008; 56: 1248-1252.

36. Abo KA, Adeyemi AA. Seasonal accumulation of anthraquinonesin leaves of cultivated *Cassia podocarpa* Guill et Perr. *Afr J Med Sci* 2002; 31: 171-173.
37. Oshio H, Kawamura N. Determination of the laxative compounds in rhubarb by high performance liquid chromatography. *Shoyakugaku Zasshi* 1985; 39: 131-138.
38. Zhou XM, Chen QH. Biochemical study of Chinese rhubarb. XXII. Inhibitory effect of anthraquinone derivatives on sodium-potassium-ATPase of rabbit renal medulla and their diuretic action. *Acta Pharmacologia Sinica* 1988; 23: 17-20.
39. Brown JP, Brown RJ. Mutagenesis by 9,10-anthraquinone derivatives and related compounds in *Salmonella typhimurium*. *Mutat Res* 1976; 40: 203-224.

Chapter 19

Evaluation of Novel Strategies for the Treatment of Aluminum-Induced Dementia in an Experimental Model

Uboweja A[a], Pandhi P[a], Malhotra S[a] and Singh AP

19.1 INTRODUCTION

Aluminum is a ubiquitous metal, which is potentially toxic to man. Aluminum (Al) accumulation has been implicated as a causative factor in a variety of disorders, abnormally high amounts of the metal have been found in various neurological conditions; including dialysis encephalopathy, amyotrophic lateral sclerosis, Down syndrome and Alzheimer's disease.[1]

Aluminum can readily cross the blood brain barrier after systemic administration and may use the same high affinity receptor ligand system that has been postulated for iron. Once in the brain, Al accumulates in various regions including the hippocampus where it can interfere with synaptic plasticity in a dose dependent manner. Application of different Al salts has generated neurofibrillary degeneration similar to that found with patients of dementia.[2]

The full scale of mechanisms underlying Al neurotoxicity probably involves multiple pathways. It has been reported that Al may interfere with neuronal signaling through interactions with glutamate receptors or calcium channels and/or intracellular calcium homeostasis.[3] Considerable evidence has been provided for interaction of Al with the cholinergic system.[4] Al was described to interfere with cholinergic transmission and

[a]*Dept of Pharamacology, Postgraduate Institute of Medical Educational and Research, Chandigarh 160012.*

signaling. The cholinergic system is also known to be particularly affected in AD and cholinergic signaling is largely involved in learning and memory mechanisms.[5]

Al can induce neuronal and glial cell death, extensive loss of synaptic contacts, and can at least potentiate the deposition of aggregated β-amyloid protein in the brain parenchyma and within the cerebro-meningeal vasculature which in turn can promote inflammatory events.[6,7] Al can also interfere with axonal transport through binding of tau protein and other neurofilament peptides and the degeneration of neurofibrils in a tangle-like conformation.[8] Advances in the understanding of both the bioinorganic chemistry of Al and the biochemistries of tau and amyloid precursor protein (APP) have strengthened the link between Al and neurofirillary tangles (NFTs) and senile plaques (SPs) from one of association to one approaching an etiology.[9]

Aluminum has long been implicated in clinical conditions like senile and presenile dementia of the Alzheimer's type. AD is the most common form of dementia in the elderly. AD is characterized histopathologically by extensive brain atrophy caused by neuron loss,[10] intraneuronal accumulation of paired helical filaments (PHFs) composed of abnormal tau proteins-neurofibrillary tangles,[11] and extracellular deposits of β-amyloid peptide (Aβ) in neuritic plaques that are surrounded by a tract of neuroinflammation in specific regions of the brain parenchyma including the cortex and hippocampus.[12] In addition to the neuropathologic lesions associated with AD, significant deficits in neurochemical functions and indices have been observed.

Treatment with cholinesterase inhibitor drugs is currently the standard of care.[13] But the average durations of treatment and beneficial effects are not optimal in all cases, because of disappointing efficacy or poor tolerability of the initial treatment as well as secondary efficacy failure or adverse effects emerging during the maintenance phase.[14] Moreover, no treatment has been shown to significantly delay the progression of the disease. Therefore, efforts to identify novel approaches in the management of patients with AD are required. This chapter attempts to detect any improvement over presently available drugs using hitherto less commonly tried therapies in AD.

Observations of the effects of *in vitro* chronic exposure of primary cultured neurons to Al have shown that some of these neurodegenerative changes can be reversed by addition of these novel agents to the culture. Hence, the present study has been designed to investigate *in vivo* the novel strategies targeting the various mechanisms implicated in the Al induced neurotoxicity leading to dementia.[15]

19.2 MATERIALS AND METHODS

19.2.1 Experimental Animals

The present study was conducted in randomly selected adult Wistar rats of either sex (150-200 g, 6 months). The animals were procured from the animal house of the institute. All the animals were housed in separate polypropylene cages (10 inch × 15 inch) containing two rats each in the departmental animal room under standard laboratory conditions of ambient temperature of 25 ± 2°C, with relative humidity of 65 ± 5 percent and a 12-dark/light cycle. All the animals were allowed standard rodent pellet and tap water *ad libitum*. Each rat was used for experimentation only once. The experiments were performed between 10.00 hr and 13.00 hr to minimize circadian influences. The Institute Animal Committee has approved the study design.

19.2.2 Drugs and Chemicals

Huperzine A was gifted by Shaanxi Jiahe Phytochem Co., Ltd., China. Rivastigmine was gifted by Torrent Pharmaceuticals Ltd., Ahmedabad. Pioglitazone, Atorvastatin were gifted by Ind-Swift Ltd., Chandigarh. Celecoxib was gifted by Unichem Laboratories Ltd., Mumbai. Glipizide was gifted by Sun Pharmaceutical Industries Ltd., Mumbai. Insulin was purchased from Knoll Pharmaceuticals Ltd., Mumbai.

19.2.3 Aluminum-Induced Dementia Model

An experimental rat model of aluminum accumulation in the brain was developed to aid in determining neurotoxity of aluminum (Al). Aluminum chloride was dissolved in distilled water to prepare a solution of concentration 10 mg/ml. Al was administered once daily by intraperitoneal injections of AIC13 (10 mg Al/kg body weight) for 30 days.

19.2.4 Experimental Model of Spatial Learning and Memory

1. Morris Water Maze

Apparatus

The apparatus consists of a circular water tank (130 cm in diameter; 50 cm in depth), filled up to a height of 30 cm with colored water maintained at 25°C. The tank was hypothetically divided into four equal quadrants and a transparent escape platform made of plexiglass (10 cm in diameter, 29 cm in height) was placed in a fixed location in the center of one of the four quadrants 1 cm below the water surface. The platform was not visible from just above the water level, and transfer trials have indicated that escape onto the platform is not achieved by visual or other proximal cues.[16] Many extra-maze cues surrounded the maze and were available for the rats to use in locating the escape platform. The platform remained fixed in this position during the training session.

Behavioral Paradigm

During habituation and all subsequent training and testing, the following conditions applied. Four cardinal points were randomly chosen as start locations, and rats were released facing the wall of the pool. The latency to find the hidden platform was recorded and used as a measure of acquisition of the task. If the rats could not locate a platform within 120 s, the trial was terminated and the rat was guided to the platform by the experimenter. After having found the platform, rats were left on it for 30 s before being placed back in their cage. A blower placed in the room was used to dry the rats after testing.

Training

Training in the maze took place over the following 5 days with one session of four trials. The platform remained in the same place during all the training sessions. Training was followed immediately by Test Session (TS1).

Testing

The procedure during all subsequent test sessions was identical to the training with an intertrial interval (ITI) of approximately 2-3 min. All parameters were counterbalanced between and within groups.

2. Passive Avoidance Task

Apparatus

A one trial step through passive avoidance task was carried out.[17] The apparatus for the step-through passive-avoidance test was divided into an illuminated compartment and a dark compartment of the same size by a wall with a guillotine door.

Adaptation Trial

In the experimental session, each mouse was trained to adapt to the step-through passive avoidance apparatus. The animal was put into the illuminated compartment, facing away from the dark compartment. After 10 s, the door between two boxes was opened and the mouse was allowed to move into the dark compartment freely. The latency to leave the illuminated compartment was recorded.

Training Trial

Two hours after the adaptation trial, the mouse was again put into the illuminated compartment. The learning trial was similar to the adaptation trial except that the door was closed automatically as soon as the mouse stepped into the dark compartment and an inescapable foot-shock (100 V, 2 s) was delivered through the grid floor.

Retention Test

The retention of passive avoidance response was measured 1, 15 and 30 days after the learning trial. Each animal was again put into the illuminated compartment and the latency to re-enter the dark compartment was recorded. No foot shock was delivered while the retention test was performed. The maximum cut off time for step through latency was 300 s.

3. Elevated Plus Maze

The elevated plus-maze test was used to evaluate spatial long term memory, following the procedure described.[18]

Apparatus

The apparatus consisted of two open (50 × 10 cm) and two closed arms (50 × 10 × 40 cm) facing each other with an open roof extending from a central platform mounted on a plywood base raised 50 cm above the floor. Light levels on the open and closed arms were similar.

Training

On the 1st day, each mouse was placed at the end of an open arm. The TL, the time taken by the mouse to move into one of the enclosed arms was recorded on the 1st day. An arm entry is defined as the entry of all the four feet of the animal into the closed arm. If the animal did not enter an enclosed arm within 90 s, it was gently pushed into one enclosed arm, and the TL was assigned as 90 s. The mouse was allowed to explore the maze for 20 s and was then returned to its home cage.

Retention Trial

Retention was examined 1, 15 and 30 days after the 1st day trial. Each mouse was again placed into the maze, and the TL was recorded. A long latency period to reach the enclosed arm indicated poor retention as compared with significantly shorter latencies.

4. Rota Rod Procedure[19]

This measured the muscle strength and coordinated movements of the animals. Rats were placed on the metallic rod (2 cm) in diameter rotating at a rate of eight revolutions per minute. Circular section divided the linear space of the rod into four lengths so that four rats could be tested together. The rats were initially screened for their ability to maintain themselves on the rotating rod for more than 3 min. If the animal after treatment could not remain on the rod for three successive trials of 3 min each, the test was considered positive i.e., motor in coordination was produced by the test compound. Rota Rod performance was evaluated as fall-off time in seconds from the rotating rod (8 rpm/min) within a period of 3 min.

All the animals were treated with aluminum chloride and were given specific treatments according to their groups (each group containing 10 animals) for the last 15 days (i.e., 16th to 30th day) of aluminum chloride treatment as follows:

A. **Control:** A separate control group was maintained under conditions similar to that of test groups except that these animals received Al for the complete study period i.e., 30 days.
B. **Vehicle:** Was maintained under similar conditions to that of the test groups except that these animals were maintained without Al and were given normal saline IP for 30 days.

Treatment Groups
Drug for last 15 days

C. **Huperzine A treated group:** Huperzine A in a dose of 0.2 mg/kg/day IP
D. **Rivastigmine treated group:** Rivastigmine in a dose of 0.4 – mg/kg/day IP
E. **Insulin treated group:** Insulin in a dose of 0.1 U/kg/day IP
F. **Glipizide treated group:** Glipizide in a dose of 20 mg/kg/day IP
G. **Pioglitazone treated group:** Pioglitazone in a dose of 10 mg/kg/day IP
H. **Atorvastatin treated group:** Atorvastatin in a dose of 10 mg/kg/day IP
I. **Celecoxib treated group:** Celecoxib in a dose of 5 mg/kg/day IP

SCHEDULE: One baseline and two more readings of the behavioral paradigms were taken at 15th and 30th day of the experiment.

STATISTICAL ANALYSIS

All data are presented as mean ±SD

The following comparison was made:

- Intergroup comparison
- Intragroup comparison

Difference between groups was calculated with One-Way ANOVA supplemented with post hoc Scheffe's test. P value ≤ 0.05 was considered to be statistically significant.

19.3 RESULTS

19.3.1 Aluminum Model of Dementia

An experimental rat model of aluminum accumulation in the brain was developed to aid in evaluating the neurobehavioral changes reminiscent of those observed in AD. Al was administered intraperitoneally (10 mg/kg/day) for 30 days to the control group. The vehicle group was given normal saline IP for the same period. Performance of both the groups was assessed at 15-day interval in all the three-neurobehavioral paradigms.

Tables 19.1, 19.2 and 19.3, give the latency values in the Morris water maze, Passive avoidance task and elevated plus maze (Mean SD). The performance of the control group animals at all the three parameters was significantly lower than the vehicle group. There was a constant decline in cognitive function over the 30-day period. The values at 15-day period show statistical significance showing that Al leads to loss of memory at this period. The values at 30-day period are highly significant in the tests performed.

Figure 19.1 compares the latency values with the control and vehicle treatment in the different tasks performed at day 1, 15 and 30. P value for the vehicle group was significant at day 15 and 30.

Aluminum model of dementia

Table 19.1 Latency values (in seconds) in neurobehavioral paradigms at Day 1 (Mean ±SD)

	Morris Water Maze	*Passive Avoidance Test*	*Elevated Plus Maze*
Vehicle	20 (7.9)	295.8 (11.7)	32.8 (10.5)
Control	17.6 (9.4)	283.3(32.5)	33.9 (11.8)

P = not significant

Table 19.2 Latency values (in seconds) in neurobehavioral paradigms at Day 15 (Mean ±SD)

	*Morris Water maze**	*Passive Avoidance Test**	*Elevated Plus Maze**
Vehicle	21.7 (7.6)	253.2 (36.9)	35.5 (13.1)
Control	49.7 (11.7)	142.5 (24.8)	51.2 (14.9)

Table 19.3 Latency values (in seconds) in neurobehavioral paradigms at Day 30 (Mean ±SD)

	Morris Water Maze#	*Passive Avoidance Test#*	*Elevated Plus Maze#*
Vehicle	23.8 (8.4)	232.5 (40.5)	44 (11.3)
Control	63.5 (14.5)	62.8 (16.9)	71.1 (13.8)

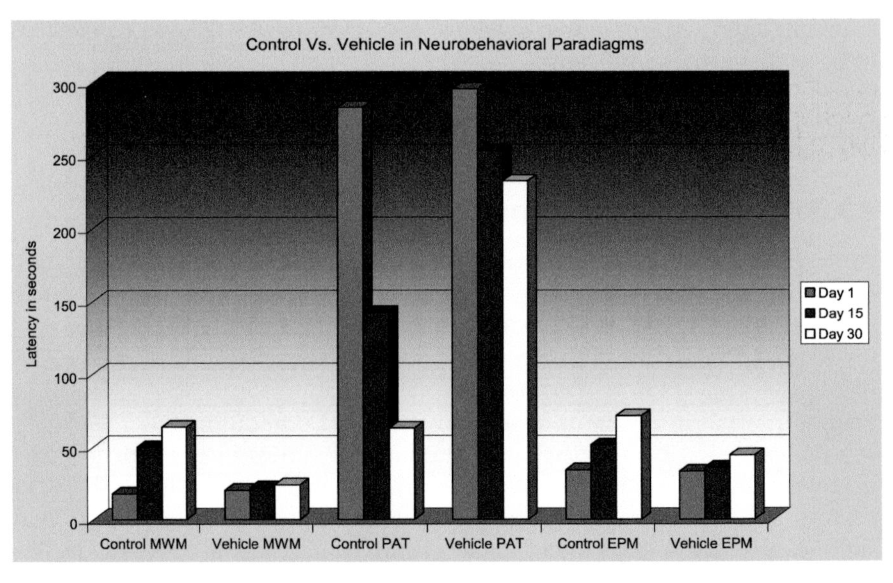

Figure 19.1 Comparison of the latency values with the control and vehicle treatment in the different tasks performed at day 1, 15 and 30.

19.3.2 Learning and Memory in Morris Water Maze Paradigm

Water Maze represents a more specific test for spatial memory, not confounded by working memory defects. The decrease in the latency to escape on the hidden platform on subsequent trials is associated with a true spatial learning and memory about the location of the platform.

The baseline retention (memory) values of escape latency among control (Al treated group for 30 days) and drug treatment groups (drug treatment during the last 15 days i.e., day 16 to day 30) are given in Table 19.4. There is no statistical difference in baseline values of all the groups (P = 36).

On treatment with Al for 15 days, there was no difference among the groups for the escape latencies as expected, since all the animals received Al treatment to cause dementia (between groups (P = 09), Table 19.5. During this period there was decline in cognitive function among the groups.

Table 19.6. shows the latency period values after drug treatment for the last 15 days. The treatment did not lead to any significant improvement in escape latencies showing that drug treatment was unable to reverse the decline in cognitive function caused by Al administration. Rivastigmine and huperzine groups though not significant to the control, have values that show some improvement in memory as compared to the other drug

Table 19.4 Latency to escape on the platform in Morris Water Maze Paradigm Day 1

Groups	*Control*	*Riva-stigmine*	*Huper-zine*	*Pioglita-zone*	*Insulin*	*Glipizide*	*Celecoxib*	*Atorva-statin*
Mean	17.6	19.3	18.7	26.4	24.4	19.2	25.6	23.5
SD	9.4	7.77	7.15	16.77	9.75	8.02	9.38	12.46

P = not significant for all comparisons

Table 19.5 Latency to escape on the platform in Morris Water Maze Paradigm Day 15

Groups	*Control*	*Riva-stigmine*	*Huper-zine*	*Pioglita-zone*	*Insulin*	*Glipizide*	*Celecoxib*	*Atorva-statin*
Mean	49.7	38.5	41.2	37.9	51.1	35.6	37.3	45.8
SD	11.68	10.76	16.56	21.81	12.86	18.65	11.56	12.32

P = not significant for all comparisons

Table 19.6 Latency to escape on the platform in Morris Water Maze Paradigm Day 30

Groups	*Control*	*Riva-stigmine*	*Huper-zine*	*Pioglita-zone*	*Insulin*	*Glipizide*	*Celecoxib*	*Atorva-statin*
Mean	63.5	45.2	47.2	54.7	64.1	56.5	55.5	58.6
SD	14.53	9.86	17.1	14.59	14.44	14.53	14.8	11.32

P = not significant for all comparisons

treated groups. Figure 19.2 shows the effect of Al and subsequent treatment with the drugs for effect on the latency values in the Morris water maze paradigm.

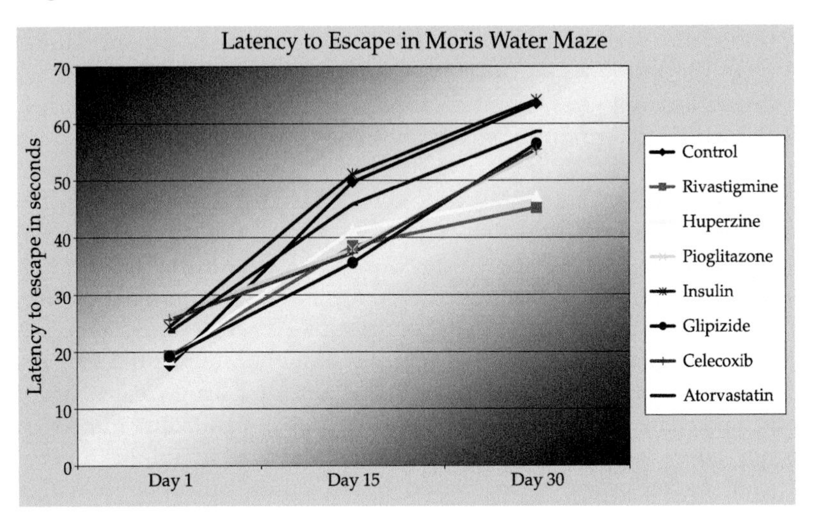

Figure 19.2 Effect of Al and subsequent treatment with the drugs for effect on the latency values in the Morris water maze paradigm.

Color image of this figure appears in the color plate section at the end of the book.

Figure 19.3 shows the comparison between the mean difference in escape latencies at the three separate readings at days 1, 15 and 30. There is an increase in escape latency among the groups at day 15 from the baseline

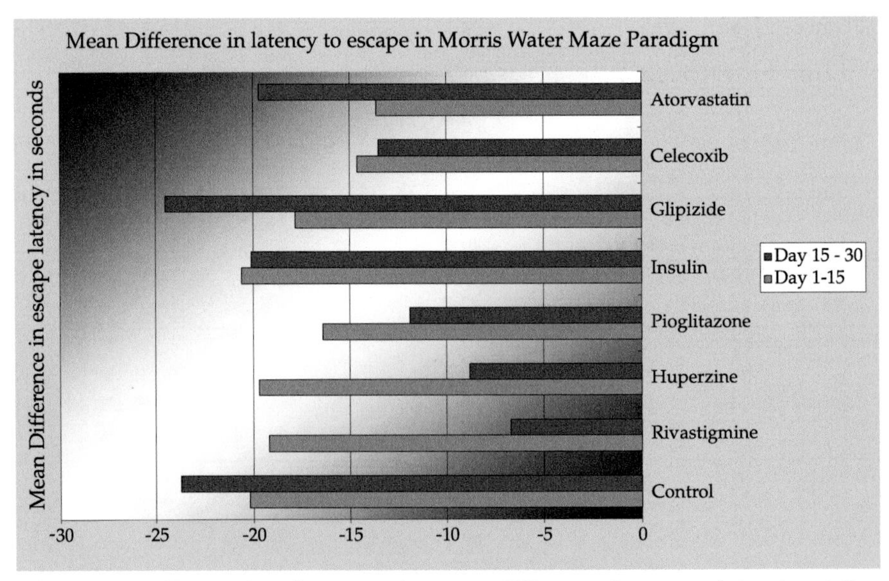

Figure 19.3 Comparison between the mean difference in escape latencies at the three separate readings at days 1, 15 and 30.

values but is not significant. This increase in escape latency (signifying loss of memory) continues further even on treatment with the drug in the last 15 days, but when compared to the control the values were not significant except the vehicle treated group (between groups P value D 15 to D 30 = 0.06).

Cognitive function assessed by passive avoidance task

It is a classic model for the assessment of cognitive performance after brain lesions or pharmacological manipulation. The latency of the animal from stimuli onset to escape after the pre-training is related to retention of memory task.

Table 19.7 shows the initial baseline retention latencies to enter the dark arm among control and treatment groups. There is no significant difference in the initial retention latencies to enter (LTE) between the groups (P = 94).

On testing for retention latency $AICl_3{\cdot}6H_2O$ treated animals showed reduced latencies compared to the vehicle group, Table 19.8. There was no significant difference among the control and drug treated groups (P = 37).

On treating Al treated rats with drugs Table 19.9 and Fig. 19.4; there was no significant improvement in LTE among the drug treated groups. Rivastigmine and huperzine treatment groups though not significant as compared to the control have improved values as compared to the other treatment groups. The vehicle has significant value (P = 001) to control showing that Al causes significant decline in cognitive function at a dose of 10 mg/kg body weight IP given for 30 days.

TABLE 19.7 Retention latency in Passive Avoidance Task Day 1

Groups	*Control*	*Riva-stigmine*	*Huper-zine*	*Pioglita-zone*	*Insulin*	*Glipizide*	*Celecoxib*	*Atorva-statin*
Mean	283.3	269.3	259.9	278.5	276.8	271	281.5	286.9
SD	32.48	64.8	57.63	47.84	19.92	53.47	51.24	41.42

P = not significant for all comparisons

Table 19.8 Retention latency in Passive Avoidance Task Day 15

Groups	*Control*	*Riva-stigmine*	*Huper-zine*	*Pioglita-zone*	*Insulin*	*Glipizide*	*Celecoxib*	*Atorva-statin*
Mean	142.5	141.6	152.3	167	172.1	142.3	181.3	179
SD	24.79	61.47	65.9	74.69	71.52	65.9	90.6	101.36

P = not significant for all comparisons

Table 19.9 Retention latency in Passive Avoidance Task Day 30

Groups	*Control*	*Riva-stigmine*	*Huper-zine*	*Pioglita-zone*	*Insulin*	*Glipizide*	*Celecoxib*	*Atorva-statin*
Mean	62.8	114.9	129.8	88.13	96	79.1	67.6	93.2
SD	16.94	62.24	63.89	47.25	46.37	77.11	52.05	84.79

P = not significant for all comparisons

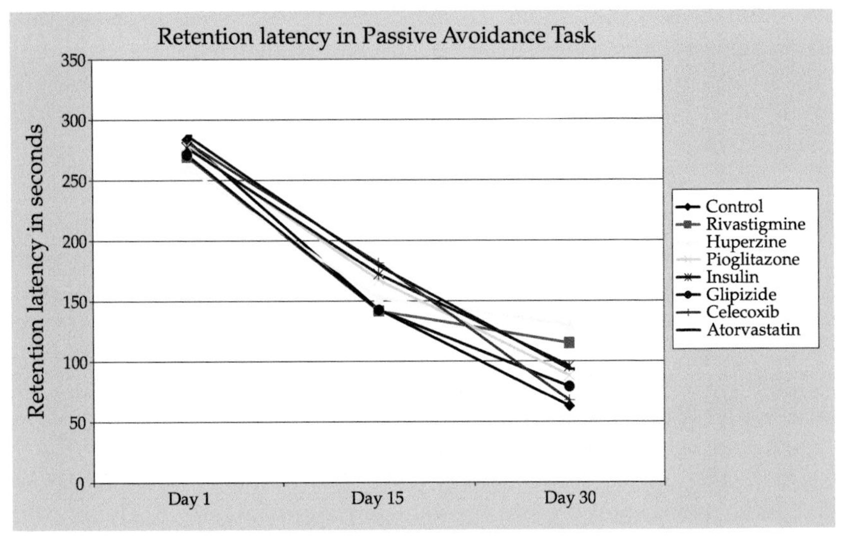

Figure 19.4 No significant improvement in LTE among the drug treated groups. Rivastigmine and huperzine treatment groups though non-significant as compared to the control have improved values as compared to the other treatment group.

Color image of this figure appears in the color plate section at the end of the book.

Figure 19.5 shows the mean difference in values between Day 1 and Day 15 among the groups. The values show the decline in retention latencies caused by Al induced decline in cognitive function. The same graph shows the difference in groups treated with the drug for 15 days to see improvement in escape latencies with the treatment. There was no

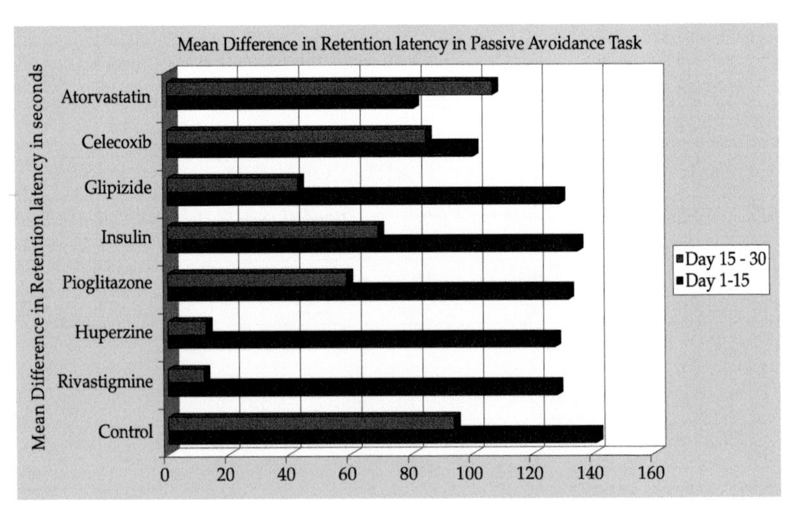

Figure 19.5 Mean difference in values between Day 1 and Day 15 among the groups. The values show the decline in retention latencies caused by Al induced decline in cognitive function.

significant improvement among the groups with the drug treatment. There was a further decline in retention latency in the control group as compared to the vehicle treated group. The P values for Di15 to 30 for the rivastigmine and huperzine are 4 and 5, respectively. These two groups show better performance in comparison to the other categories.

19.3.3 Elevated Plus Maze

There was no significant difference in transfer latencies among the groups (control vs. drug treatment) at the baseline, Day 1 (Table 19.10).

Table 19.10 Transfer latency to enter the dark arm in Elevated plus Maze Day 1

Groups	*Control*	*Riva-stigmine*	*Huper-zine*	*Pioglita-zone*	*Insulin*	*Glipizide*	*Celecoxib*	*Atorva-statin*
Mean	33.9	35.5	39.5	38.4	34.3	40.1	37.9	41.7
SD	11.81	21.44	18.48	29.09	26.4	11.75	16.2	15.4

P = not significant for all comparisons

Transfer latency in Al treated group increased, although there is no significant difference among the groups at day 15 (between groups P = 55) (Table 19.11).

Table 19.11 Transfer latency to enter the dark arm in Elevated plus Maze Day 15

Groups	*Control*	*Riva-stigmine*	*Huper-zine*	*Pioglita-zone*	*Insulin*	*Glipizide*	*Celecoxib*	*Atorva-statin*
Mean	51.2	48.8	52.1	56.9	52.3	59.3	49.6	58.3
SD	14.85	18.6	16.48	25.65	24.53	10.36	15.5	15.7

P = not significant for all comparisons

Al treatment for 30 days produced a significant increase in latency which was not reversed by the drug treatment, showing that the drugs are unable to correct the decline in cognitive function caused by aluminum (Table 19.12 and Fig. 19.6).

Table 19.12 Transfer latency to enter the dark arm in Elevated plus Maze Day 30

Groups	*Control*	*Riva-stigmine*	*Huper-zine*	*Pioglita-zone*	*Insulin*	*Glipizide*	*Celecoxib*	*Atorva-statin*
Mean	71.7	54.8	60.3	68.6	66.4	70.8	65.3	69
SD	13.83	11.9	12.39	19.67	18.16	19.05	16.19	12.67

P = not significant for all comparisons

The mean differences in values among the groups are given in Tables 19.13 to 19.15. The non-significant difference at day 15 signifies that all the groups are similar in cognitive function before the start of the drug treatment. The difference between day 15 and 30 shows if there is any improvement with the drug treatment. The decline in escape latency values is lesser with the drug treatment in almost all the treatment groups, rivastigmine and huperzine show better performance after the treatment among all the treatment groups, this values are not statistically significant, Fig. 19.7.

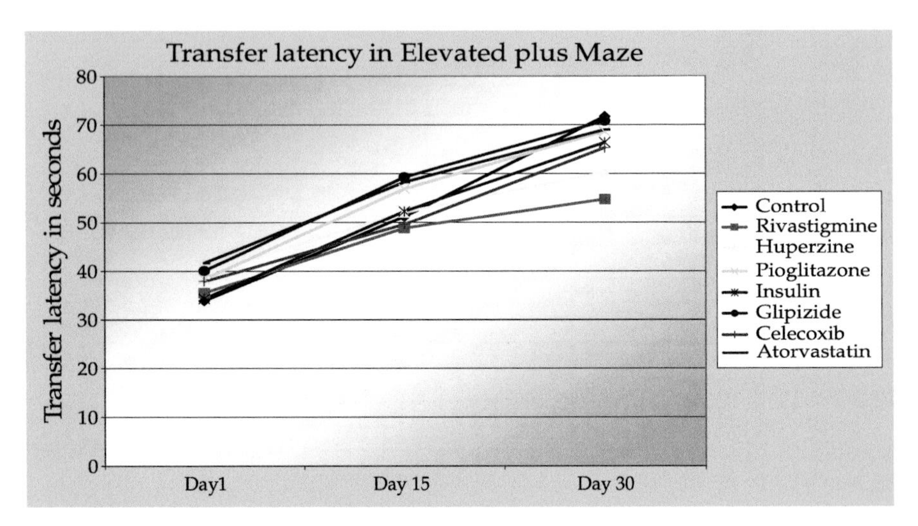

Figure 19.6 Significant increase in latency which was not reversed by the drug treatment, showing that the drugs are unable to correct the decline in cognitive function caused by aluminum.

Color image of this figure appears in the color plate section at the end of the book.

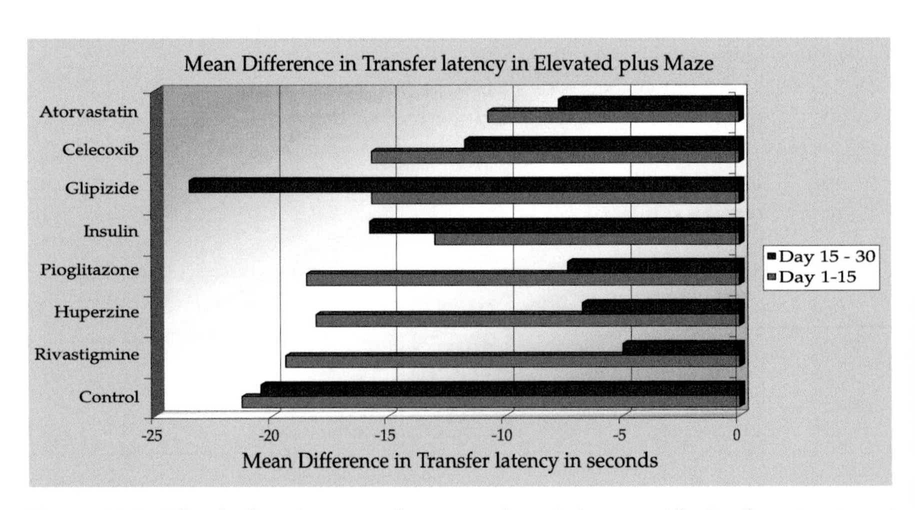

Figure 19.7 The decline in escape latency values is lesser with the drug treatment in almost all the treatment groups, rivastigmine and huperzine show better performance after the treatment among all the treatment groups, this value is not statistically significant.

Table 19.13 Mean Difference in Transfer latency in Elevated plus Maze between Day 1 and Day 15 (Day 1-Day 15)

Groups	*Control*	*Riva-stigmine*	*Huper-zine*	*Pioglita-zone*	*Insulin*	*Glipizide*	*Celecoxib*	*Atorva-statin*
Mean	–21.3	–19.4	–18.1	–18.495	–13	–15.7	–15.7	–10.7
SD	11.85	19.65	29.43	30.69	29.13	8.89	19.71	15.84

P = not significant for all comparisons

Table 19.14 Mean Difference in Transfer latency in Elevated plus Maze between Day 15 and Day 30 (Day 15-Day 30)

Groups	*Control*	*Riva-stigmine*	*Huper-zine*	*Pioglita-zone*	*Insulin*	*Glipizide*	*Celecoxib*	*Atorva-statin*
Mean	–20.5	–5	–6.7	–7.333	–15.8	–23.5	–11.7	–7.7
SD	16.02	12.02	24.17	28.56	20.27	17.36	23.75	21.84

P = not significant for all comparisons

Table 19.15 Mean Difference in Transfer latency in Elevated plus Maze between Day 1 and Day 30 (Day 1-Day 30)

Groups	*Control*	*Riva-stigmine*	*Huper-zine*	*Pioglita-zone*	*Insulin*	*Glipizide*	*Celecoxib*	*Atorva-statin*
Mean	–41.8	–24.4	–24.8	–25.828	–28.8	–39.2	–27.4	–18.4
SD	18.29	22.07	27.69	32.79	26.23	21.53	26.04	24.69

P = not significant for all comparisons

19.4 DISCUSSION

In this chapter, drugs from various categories namely anti-inflammatory agents, insulin sensitizing agents, lipid lowering agents and herbal medicine and cholinerterase inhibitors were tested in the aluminum model of dementia. Several interesting findings were observed in the study and the effect of these drugs on each parameter were discussed step by step.

Behavior can be defined as the end product of a variety of sensory, motor and integrative processes occurring in the nervous system.[20] The functional capacity of the central nervous system cannot be determined by histological or even physiological studies independent of behavioral analysis.[21]

Neurobehavioral methods are being used with increasing frequency in toxicity studies to assess the deleterious effects of chemicals and physical factors on the presumption that they are more sensitive than other tests in determining toxicity due to the fact that behavior is a functional indicator of the net sensory, motor and integrative processes occurring in the central and peripheral nervous system.

Aluminum chloride affect the nervous system and produce effects which are manifested in the form of a variety of symptoms and no single method/test can evaluate its neurobehavioral toxicity. The present investigation includes a battery of tests to evaluate neurobehavioral effects of $AlCl_3$ administered intraperitoneally to the rats to cause dementia.

19.4.1 Neurobehavioral Effects of Aluminum Chloride

The neurobehavioral toxicity has been evaluated by using a battery of neuropsychobehavioral tests evaluating all aspects of memory, such as, spatial memory, conditioned response and working memory. The paradigms used in the study are Morris water Maze, Elevated plus Maze Passive avoidance test, rota rod test.

Spatial memory (a type of declarative memory) is evaluated by various models of which Morris water maze is considered to be the best model due to several advantages 1) Food and water deprivation is not required in this test, 2) water provides a uniform intra maze environment, thus eliminating any olfactory interference. In various experiments Morris water maze has been successfully used for evaluation of anti-dementia and anti-amnesic drugs.

The conditioning processes have been considered to be the basic element of learning.[22] The ability to learn and remember has as adaptive value for an animal. The ability to learn allows an animal to escape or avoid aversive situations or, approach desirable objects and to store the memory of these contingencies for future use. However, the specific result of a toxic effect on learning must be interpreted within the limitations of the experimental design and in conjunction with controls that assess drug-induced changes on sensory, motor and motivational processes. One-way shock avoidance tasks require the animal to move from one compartment to another in order to avoid or escape the shock.[23] This test has been included to evaluate the effect of Al salt on learning and memory.

A study on the effects of drugs on the motor activity is one of the important parameters in the neuropsychopharmacological investigations. The rota rod test has been devised as a simple and sensitive quantitative test for evaluation of muscle strength and motor co-ordination. Elevated plus maze was used for evaluation of anxiety. Nowadays it is also being used for evaluating learning and memory.

19.4.2 Al Model of Dementia

An aluminum-induced dementia model is a validated model of memory. Various studies have been done evaluating the aluminum-induced neurobehavioral effects and morphological changes in the rat brain.[24,25] Animals loaded with aluminum develop both symptoms and brain lesions that are similar to those found in AD.

The performance of the control group animals (Al treatment for 30 days) at all the three-neurobehavioral paradigms was significantly lower than the vehicle group (normal saline treatment for 30 days). The decline in cognitive function was assessed at 15th day and 30th day of treatment. There was a constant decline in memory of Al treated animals.

19.4.3 Cholinesterase Inhibitors and Dementia (Positive Control Group)

Efficacy of cholinesterase inhibitors is modest.[26] They have no effect on disease progression and use for more than a year, has been largely unstudied. These are some important reasons why we have tried to find other strategies which have more benefits for treating dementia.

In this present study, rivastigmine showed modest benefits as shown by a lesser decline in cognitive function in comparison to the control, while the difference was not statistically significant, the performance of animals was much better in comparison to the other novel strategies tried.

Insulin, glucose and memory

Studies have confirmed that glucose administration can facilitate memory in healthy humans and in patients with Alzheimer's disease. Interestingly, glucose effects on memory appear to be modulated by insulin sensitivity (efficiency of insulin mediated glucose disposal).[27] Biological data suggests that insulin may contribute to normal cognitive functioning and that insulin abnormalities may exacerbate cognitive impairments.

In animals, systemic insulin administration has been associated with memory deficits, likely due, in part to hypoglycemia that occurs when exogenous insulin is not supplemented with glucose to maintain euglycemia. Therefore, in this study insulin administration was preceded by fructose administration by half an hour to prevent peripheral hypoglycemia. In the present investigation, the dose of insulin with fructose did not improve memory significantly even though the values are somewhat better than the control.

Results of a recent study indicate a direct action of prolonged (8 week) intranasal administration of insulin or brain functions, improving declarative memory and mood in the absence of systemic side effects.[28] Many studies are available which have evaluated the effect of insulin on the neuropathological mechanism of memory deterioration,[29] while a few studies have evaluated the effect of insulin treatment on retrieval of memory for a long duration of time. Studies, which have shown a positive effect of insulin administration of learning and retrieval, are short duration studies.

Epidemiological evidence suggests that insulin resistance influences the risk of developing AD.[30,31] The increasing evidence of insulin resistance in AD and other numerous mechanisms through which insulin may affect clinical and pathological aspects of the disease suggests that improving insulin effectiveness may have therapeutic benefits for patients of AD. However, it is well known that data from observational studies may not be replicated in humans, it has been shown that even experimental evidence does not support this strategy.[32]

In this study we observed the effect of Pioglitazone, an insulin sensitizing agent in reducing the cognitive decline with AI treatment. The results demonstrate that there is no significant benefit of treatment on the loss of memory function with the drug treatment. In an earlier study using neuronal cultures and *in vivo* model of neuro-inflammation, it was seen that troglitazone antagonized the Aβ stimulated pro-inflammatory response and neurotoxicity. Neurobehavioral animal studies have not assessed the role of these agents in the cognitive function, even though many *in vitro* studies are available which have assessed the effect of these agents on various pathophysiological mechanisms implicated in AD9.[28] Here too the decline in cognitive function with the pioglitazone arm was less in comparison to the control group, however this does not reach statistical significance.

Oral hypoglycemic agents have not been evaluated to test for acquisition or retrieval of memory. In fact in one study, tolbutamide was shown to produce memory deficits, as tolbutamide displaces A from albumin, which increases the concentration of free A and enhances amyloid deposition in an *in vitro* assay.[33] In this study we analyzed the effect of Glipizide on the learning memory. The results in the behavioral paradigms are not significantly different from the control group.

None of the medication affecting insulin levels or sensitivity was significantly effective in improving the deterioration in memory.

Anti-inflammatory agents and memory

Some epidemiological and pilot clinical studies have proven that long-term administration of anti-inflammatory drugs has a protective effect on the onset of AD.[34] Among them; non-steroidal anti-inflammatory drugs (NSAIDs) are the most extensively investigated medicaments. Despite some contradictory findings, the prevalent majority of these studies prove that long-term application of anti-inflammatory treatment can delay the onset, or at least slow down the progression of AD, namely in people between 65 and 75 yr of age. The most appropriate prophylactic effect seems to be achieved by specific inhibitors of cyclooxygenase-2 (COX-2).

In our study, we did not find any effect of selective COX-2 inhibitor, Celecoxib on memory function in all the three memory tasks ($P = 9$). This is in contradiction to the study[35] which showed significant improvement in open field behavior ($P < 0.05$) in mice treated with ibuprofen. Another study[36] has shown that the retention test escape latencies of rats administered indomethacin (5 and 10 mg/kg) or NS-398 (2 and 5 mg/kg) were significantly higher than those of vehicle–treated rats in the hidden platform task, indicating an impairment in retention. This shows that the results with the agent are equivocal.

A meta-analysis of epidemiological studies has shown the pooled relative risk of Alzheimer's disease among users of NSAIDs was 0.72 (95 percent confidence interval 0.56 to 0.94).[37] But these results have not been supported by a recently conducted randomized clinical trial, which reported the failure of selective cyclo-oxygenase-2 to slow the progression of AD.[38]

Statins and Dementia

There is increasing data from epidemiological studies and from model systems indicating that cholesterol-lowering drugs may have an impact on the development of memory deterioration in AD.[39,40] The agents have been shown to be neuroprotective and restore memory loss caused by TBI (traumatic brain injury) in animal models. In a study conducted by Lu et al., atorvastatin promoted the restoration of spatial memory function in an experimental model of TBI (traumatic brain injury).[41] The role of these agents on cognitive parameters has not been assessed in models of dementia; although studies are available which show their modulatory effects on the biochemical parameters involved in the pathogenesis of dementia in Alzheimer's disease.

In our study we were unable to find any significant advantage in neurobehavioral parameters of animals treated with atorvastatin as assessed in various tasks. As compared to the other treatment groups the performance of animals in the behavioral tasks did not deteriorate with the treatment. In fact the values are quite near to the Rivastigmine and Huperzine group but there is no statistical significance among the groups.

Huperzine, herbal medicine and dementia

Huperzine A is a novel alkaloid isolated from the Chinese herb *Huperzia serrata* Thunb. Pron. (Lycopodiaceae). Dementia and cognitive impairment in Alzheimer's disease have been attributed to the hypofunction of cholinergic neurotransmission in the brain. Previous studies demonstrate that Huperzine A is a reversible and selective inhibitor of acetyl choinersterase.[42]

Numerous studies, most of them from China, suggest that huperzine A may be as effective as the drugs tacrine and donepezil in AD.[43,44] In other countries, it is available as a nutriceutic and is being used by some clinicians to treat AD. There have been no controlled clinical trials outside China assessing the toxicity and efficacy of these agents and comparing them with acetyl cholinerterase.

In the present study, we have seen some improvement in the cognitive function with the herb in comparison to the other drug treatment arms. As per evidence available, the effect with Huperzine was comparable to the positive control, Rivastigmine. The performance in various tasks of memory though was statistically significant with the control group. These are the type of results we expect to get in actual scenarios, as these agents are more beneficial for symptomatic improvement and have less effect on modifying the progression of the disease.

The findings of our study suggest that the novel agents being implicated for the prevention or treatment of dementia in AD are of not much benefit. The results are somewhat better than the control but do not reach any significance and it is not expected that it will be so in clinical practice.

19.5 CONCLUSIONS

Children argue about what's scarier: ghosts or monsters? Fire or sharks? Adults aren't above ideally comparing their own fears. Which would be worse: cancer or heart disease? A car crash or AIDS? But a strong case can be made that the scariest thing about growing old is the fear of losing one's mind through Alzheimer's disease. There is a sense of urgency and optimism among researchers about possible treatments and perhaps prevention of the disease. A few of these therapies are already in use (cholinesterase inhibitors, NMDA antagonist), or in clinical trials (COX-2) inhibitors, insulin sensitizing agents), even more are at a proof of concept stage.

Several experimental studies have shown the efficacy of some of these novel therapies. However, most of them have observed the effects of these drugs on biochemical parameters; have been conducted for a shorter duration of time and using other models. There is not much information on the effectiveness of these agents on the cognitive function over a long period of time.

Therefore, we investigated the effects of novel agents on memory in the aluminum model of dementia.

The results of this study indicate that:

1. One general trend was observed in the study—all the values observed in the three-neurobehavioral paradigms with most of the treatment groups were better than the control, though no statistical significance was observed. This could not be a chance finding as neurobehavioral tasks were performed. Such results are expected in neurological studies, which are more of a subjective system where the placebo response is quite high, but that this cannot possibly be evaluated in animals.
2. The drugs did not demonstrate significant efficacy in Morris water Maze paradigm.
3. The novel agents proposed were no different from the control in Passive Avoidance task either.
4. In the elevated plus maze paradigm, no significant effect of the agents on reversal of cognitive decline was demonstrated. The study demonstrates no significant benefits with the anti-inflammatory agents, lipid lowering drugs, anti-diabetic agents in the animal model of dementia.

Several questions remain unanswered. First, whether these agents can be used as a preventive therapy being implicated by epidemiological studies. Secondly, randomized controlled clinical studies are required to elucidate the efficacy of these agents in patients with dementia. Or, better still, novel approaches to manage this distressing disorder may be sought and evaluated.

REFERENCES

1. Zatta PF. Biological models for the study of aluminium neurotoxicity. *Acta Medica Romanica* 1997; 35: 592-600.
2. Klatzo I, Wiesniewski H, Streciher E. Experimental production of neurofibrillarty degeneration: Light microscopic observation. *J Neuropath Experi Neuro* 1965; 24: 187-189.
3. Platt B, Fiddler G, Riedel G, *et al.* Aluminium Toxicity in the rat brain: Histochemical and immunohistochemical evidence. *Brain Research Bulletin* 2001; 55: 257-267.
4. Meiri H, Banin E. Roll M, *et al.* Toxic effect of Aluminium on nerve cell and synaptic transmission. *Prog Neurobiol* 1993; 40: 89-121.
5. Cain DP. Testing the NMDA long-term potentiation and cholinergic hypotheses of spatial learning. *Neurosc Behav Rev* 1998; 22: 181-193.
6. Suarez-Fernandz MB, Solado AB, Sanz-Mendel A. Aluminium induced degeneration of astrocytes occurring via apoptosis and results in neuronal death. *Brain Research* 1999; 835: 125-126.
7. Bondy SC, Troung A. Potentiation of beta-folding of ß-amyloid peptide by aluminium salts. *Neurosc Lett* 1999; 267: 25-28.
8. Tronsco JC, Stenberger NH, Stenberger LA, *et al.* Immunocytochemical studies of neurofilamental antigens in the neurofibrillary pathology induced by aluminium. *Brain Res* 1986; 364: 295-300.
9. Drewes G, Ebneth A, Mandelkow EM. MAPSs, MARKs and microtubules dynamics. *TIBS* 1998; 23: 307-311.
10. Whitehouse PJ, Price DL, Struble RG, *et al.* Alzheimer's disease and senile dementia: loss of neurons in basal forebrain. *Science* 1982; 215: 1237-1239.
11. Holzer M, Holzapfel HP, Zedlick D, *et al.* Abnormally phosphorylated tau protein in Alzheimer's disease: heterogeneity of individual regional distribution and relationship to clinical severity. *Neuroscience* 1994; 63: 499-516.
12. Hsaio K, Chapman P, Nilsen S, *et al.* Correlative memory deficits, Abeta elevation, and amyloid plaques in transgenic mice. *Science* 1996; 4: 99-102.
13. Grutzendler J, Morris JC. Cholinesterase inhibitors for Alzheimer's disease. *Drugs* 2001; 61: 41-52.
14. Emre M. Switching cholinesterase inhibitors in patients with Alzheimer's disease. *Int J Clin Pract Suppl* 2002; 127: 61-72.
15. Ogasawara Y, Sakamoto T, Ishii K, *et al.* Effects of administration routes and chemical forms of aluminum on aluminum accumulation in rat brain. *Biol Trace Elem Res* 2002; 86: 269-78.
16. Morris R. Development of a water-maze procedure for studying spatial learning in the rat. *J Neurosci Methods* 1984; 11: 47-60.

17. Suzuki M, Yamaguchi T, Ozawa Y, *et al.* Effects of (–)-S-2-8-tartrate monohydrate (YM796), a novel muscarinic agonist, on disturbance of passive avoidance learning behaviour in drug treated and senescence accelerated mice. *J Pharm Exp Ther* 1995; 275: 728-736.
18. Pellow S, Chopin P, File SE, *et al.* Validation of open: closed arm entries in an elevated plus maze as a measure of anxiety in rat. *J Neurosci Meth* 1985; 14: 149-167.
19. Dunham NW, Miya TS. A note on a simple apparatus for detecting neurological deficits in rats and mice. *J Am Pharm Assoc* 19957; 46: 208-211.
20. Tilson HA, Harry GJ. *Behavioural principles for the use in behavioural toxicology and pharmacology In: Nervous System Toxicology.* C. Mitchell (Ed.), New York, Raven Press, 1982; pp. 1-27.
21. Mello NK. Behavioral toxicology: a developing discipline. *Fed Proc Am Sc Exp Biol* 1975; 34: 1832-1834.
22. Valzelli L. *Psychopharmacology—An introduction to experimental and clinical principles.* New York, Raven Press, 1973b; pp. 59-76.
23. Tilson, HA, Mitchell CL. Neurobehavioural techniques to assess the effects of chemicals on the nervous system. *Ann Rev Pharmacol Toxicol* 1984; 24: 425-450.
24. Platt B, Fiddler G, Riedel G, *et al.* Aluminium Toxicity in the rat brain: Histochemical and immunohistochemical evidence. *Brain Research Bulletin* 2001; 55: 257-267.
25. Roloff EL, Platt B, Riedel G. Long-term stud of chronic oral aluminium exposure and spatial working memory in rats. *Behavioural Neurosci* 2002; 116: 351-356.
26. Courtney C, Farrell D, Gray R, *et al.* Long-term donepezil treatment in 565 patients with Alzheimer's disease (AD2000): randomized double-blind trial. *Lancet* 2004; 363: 2105-2115.
27. Waston GS, Craft S. Modulation of memory by insulin and glucose: neuropsycological observation in Alzheimer's disease. *Eur J Pharmacol* 2004; 19: 97-113.
28. Benedict C, Hallschmid M, Hatke A, *et al.* Intranasal insulin improves memory in humans. *Psychoneuroendocrinology* 2004; 29: 1326-1334.
29. Carro E, Torres-Aleman I. The role of insulin and insulin-Like growth factor in the molecular and cellular mechanisms underlying the pathology of Alzheimer's disease. *Eur J Pharmacol* 2004; 490: 127-33.
30. Leibson CL, Rocca WA, Hanson VA, *et al.* Risk of dementia among persons with diabetes mellitus: a population based cohort. *Am J Epidemiol* 1997; 145: 301-308.
31. Ott A, Stolk RP, Hofman A, *et al.* Association of diabetes mellitus and dementia: the Rotterdam study. *Diabetologia* 1996; 39: 1392-1397.

32. Malhotra S, Kondal A, Shafiq N, *et al.* A comparison of observational studies and controlled trials of heparin in ulcerative colitis. *Int J Clin Pharmacol Ther* 2004; 39: 19-24.
33. Bohrmann B. Endogenous proteins controlling amyloid beta-peptide polymerization. *J Biol Chem* 1999; 274: 15990-15995.
34. McGeer PL, Schulzer M, McGeer EG. Arthritis and anti-inflammatory agents as possible protective factors for Alzheimer's disease: a review of 17 epidemiologic studies. *Neurology* 1996; 47: 425-432.
35. Lim GP, Yang F, Chu T, *et al.* Ibuprofen effects on Alzheimer pathology and open field activity in APPsw transgenic mice. *Neurobiol Aging* 2001; 22: 983-991.
36. Theather LA, Packard MG, Bazan NG. Post-training cyclooxygenase-2(COX-2) inhibition impairs memory consolidation *Learn Mem* 2002; 9: 41-47.
37. Etminan M, Gill S, Samii A. Effect of non-steroidal anti-inflammatory drugs on risk of Alzheimer's disease: systematic review and meta-analysis of observational studies. *BMJ* 2003; 327: 128-133.
38. Reines SA, Block GA, Morris JC, *et al.* Rofecoxib: no effect on Alzheimer's disease in a 1-year, randomized, blinded, controlled study. *Neurology* 2004; 62: 66-71.
39. Wolozin B, Kellman W, Ruosseau P, *et al.* Decreased prevalence of Alzheimer disease associated with 3-hydroxy-3-methylglutaryl co-enzyme A reductase inhibitors. *Arch Neurol* 2000; 576: 1439-1443.
40. Jick H, Zornberg GL, Jick SS, *et al.* Statins and the risk of dementia. *Lancet* 2001; 356: 1627-1631.
41. Lu D, Mahmood A, Goussev A, *et al.* Atorvastatin reduction of intravascular thrombosis, increase in cerebral microvascular patency and integrity, and enhancement of spatial learning in rats subjected to traumatic brain injury. *J Neurosurg* 2004; 101: 813-821.
42. Jiang H, Luo X, Bai D. Progress in clinical, pharmacological, chemical and structural biological studies of huperzine A: a drug of traditional Chinese medicine origin for the treatment of Alzheimer's disease. *Curr Med Chem* 2003; 10: 2231-2252.
43. Xu SS, Cai ZY, Qu ZW, *et al.* Efficacy of tablet huperzine-A on memeory, cognition and behavoiur in Alzheimer's disease. *Chung Kuo Yao Li Hsueh Pao* 1995; 16: 391-395.
44. Qian BC, Wang M, Zhou ZF, *et al.* Pharmacokinetics of tablet huperzine-A in six volunteers. *Chung Kuo Yao Li Hsueh Pao* 1995; 16: 396-398.

Chapter 20

Acne and Natural Products

20.1 INTRODUCTION

Acne vulgaris is a common skin disorder, that affects approximately 70-87 percent of adolescents and young adults.[1] The pathogenesis of acne is multifactorial and complex and is thought to be due to, androgen-stimulated sebum production, which leads to follicular occlusion, hyperkeritinization and over growth of bacteria; Propionibacterium acnes and Staphylococcus epidermis, leading to comedones and inflammatory papules and pustules. Conventional treatments for acne include topical exfoliating agents such as Benzoyl peroxides, retinoids, and antibiotics (topical and systemic) but symptoms may not always improve or patients may have adverse reactions to conventional treatments and thus seek alternative treatments. Antibiotic resistance in Propionibacterium acnes and Staphylococcus epidermis has also been rising, thus promoting the need to look at alternative therapies.[2]

20.2 DIETARY AND LIFESTYLE FACTORS

Whether dietary factors influence acne has been a debate for decades. One review of the literature looking at evidence for diet, face-washing and sunlight exposure in acne management concluded that the evidence is incomplete at best.[3] Another review did not support any link between acne and foods such as dairy products, chocolate, and fatty foods.[4] However, with the more recent focus on diet and nutritional supplements, emerging research is suggesting that diet may be a factor, particularly in mediating the inflammation and oxidative stress of the acne process. Western diets, with characteristically high glycemic indexes (GI), can elevate insulin and IGF-1 levels acutely and chronically.[5]

These hormones stimulate adrenal and gonadal androgen production, leading to increased sebum production and acne. Frequent consumption of high glycemic index (GI) carbohydrates may repeatedly expose adolescents to acute hyperinsulinemia. Therefore, a low glycemic load diet may have a beneficial effect on acne.[6]

A review article by Berra et al., also supported a possible correlation between a high glycemic diet and acne and suggested an improvement in acne after glycemic indexes and glycemic load were reduced.[7]

A randomized controlled trial of 43 males aged 15-25 yr showed that a low-glycemic-load (LGL) diet improved acne lesions and reduced weight and BMI. Weight loss is known to decrease circulating androgen and insulin levels; thus, it was unclear if improvement in acne was due to dietary differences or weight reduction or both.[8]

A cross-sectional, self-report study of 47,355 nurses revealed that intake of milk during adolescence was associated with history of teenage acne.[9] The authors also prospectively examined the effects of milk intake and acne in the children of these nurses and found that higher milk consumption, regardless of fat content, was associated with acne.[10,11]

The authors speculated that milk (whole or nonfat) contains hormones and bioactive molecules, such as androgens, progesterone, and insulin growth factor-1 (IGF-1), that can have an acne-stimulating effect.[12]

These cohort studies, however, can only suggest correlation but not causation. Stress has also been blamed as a trigger for acne flares. Two independent groups of researchers studied high school and university students and found that increased stress levels during examination periods, correlated with increased acne severity.[13,14]

20.3 NUTRITIONAL SUPPLEMENTS AND PHYTOCHEMICALS

Arachidonic acid, (ω-6 fatty acid, which is a major part of the western diet) is a precursor to the manufacture of pro-inflammatory leukotriene B4 (LTB4). The involvement of pro-inflammatory leukotriene B4 (LTB4) in the pathogenesis of acne has recently been described. A study looking at the administration of a novel LTB4 blocker led to a 70 percent reduction in inflammatory acne lesions.[15]

The anti-inflammatory properties of ω-3 essential fatty acids, (ω-3 EFA's) including LTB4 inhibition, are well known.[16] To date there are no studies looking at the direct effect of supplementation with ω-3 EFA's and acne but a diet high in ω-3 EFA's could have a synergistic effect with a low glycemic load diet on improving acne.[6,17]

Lactoferrin (a whey milk protein/iron-binding protein) a dietary supplement, has also been shown to decrease skin inflammation due to its broad antibacterial and anti-inflammatory activities.[18]

Two studies looked at the efficacy and tolerability of lactoferrin in adolescents and young adults with mild to moderate facial acne vulgaris and found a significant reduction in inflammatory acne lesion count in the lactoferrin group. The results suggest that dietary supplementation with bovine lactoferrin in mild to moderate acne vulgaris can decrease acne lesions.[19,20]

Another nutritional supplement called APC (methionine-based zinc complex, chromium, and vitamins with anti-oxidants) was studied in a pilot study of 48 patients (15-35 yr) with acne. Oral APC was given thrice daily for 3 months followed by a 4-week treatment-free period. At the end of the treatment (Week 12), there was a statistically significant improvement in the global acne count with decrease in pustules and papules ($P < 0.001$).[21]

Resveratrol is a natural compound produced by some spermatophytes such as grapes and other plants and has been shown to be anti-inflammatory and active against several bacteria including Propionibacterium acnes.[22]

A single-blind pilot study of 20 patients looked at the therapeutic effects of resveratrol on acne vulgaris. Resveratrol gel was applied daily on the right side of the face for 60 days, compared to a control hydrogel on the left side of the face. There was a 53.75 percent mean reduction in the Global Acne Grading System score on the resveratrol-treated sides of the face compared with 6.10 percent on the control side.[23]

20.4 BIOCHEMICAL THERAPIES

Few herbal medicines have been evaluated systematically in clinical trials. Witch hazel bark (*Hamamelis virginiana*) has been used for acne because of its naturally astringent properties.[24] However to date no randomized trials have been done to substantiate its use, but and is often used in acne products.

Green tea extract and tea tree oil have been investigated in the treatment of acne. Tea Tree oil is an essential oil of the Australian native tree, *Melaleuca alternifolia* has been shown to have antibacterial and antifungal properties.[25]

A randomized, double-blind placebo-controlled study investigated the efficacy of 5 percent topical tea tree oil gel in 60 patients (15 to 25 yr) with mild to moderate acne vulgaris. Tea tree oil gel was 3.55 times and 5.75 times more effective than the placebo.[26]

A single-blind, randomized clinical trial of 124 patients with mild to moderate acne evaluated the efficacy and skin tolerance of 5 percent tea-tree oil (Melaleuca alternifolia) with 5 percent benzoyl peroxide lotion. After 3 months of treatment, both 5 percent tea-tree oil and 5 percent benzoyl peroxide significantly reduced acne but fewer side effects were reported with the use of tea tree oil (44 vs 79 percent) although the onset of action was slower in tree tea oil.[27]

Green tea has catechins which are phytochemical phenolic compound that have anti-inflammatory effects. Two studies looked at the efficacy of topical 2 percent green tea lotion (natural plant extract) in the treatment of mild-to-moderate acne vulgaris in adolescents and young adults. Topical 2 percent green tea lotion was significantly effective for mild-to-moderate acne vulgaris.[26,28,29]

Several other natural ingredients such as colloidal oatmeal, feverfew, licorice, aloe vera, chamomile, curcumin, soy, coffeeberry, mushroom extracts, green tea, pine bark extract, vitamin E, vitamin C, and niacinamide have anti-oxidant, anti-inflammatory, and/or moisturizing properties thus these ingredients may be effective adjuncts with other acne therapies to decrease the erythema associated with inflammatory acne.[30]

20.5 AYURVEDIC HERBS

Two randomized, double-blind, placebo controlled clinical studies exploring the efficacy of Ayurvedic treatment in acne have been published. These studies indicate that some Ayurvedic remedies might be effective for acne. One study demonstrated that the combination of an oral Ayurvedic preparation and a topical Ayurvedic, multicomponent preparation (cream or gel) was more efficacious in treating acne than oral therapy alone.[31]

Another study showed treatment with an oral preparation of sunder vati resulted in a significant reduction in acne lesion count and was well tolerated. Treatment with the three other oral formulations studied failed to show any improvement.[32]

20.6 TRADITIONAL CHINESE MEDICINE/ ACUPUNCTURE

A few small studies in adults have looked at the effectiveness of Traditional Chinese herbs and acupuncture for treatment of acne and found them to be promising. A double-blind controlled trial evaluated the efficacy of ah shi point and general acupuncture point treatment of acne vulgaris in 36 adults. Ah shi point acupuncture involves inserting a needle at painful or pathological sites (papules and nodules of acne) to reduce inflammation of the acne site directly. After 12 treatment sessions, there was a significant reduction in the inflammatory acne lesion counts, the quality-of-life scale Skindex-29 scores and the subjective symptom scores from baseline in both groups.[33]

A Chinese herbal compound, Compound Oldenlandis Mixture was compared with Angelica and Sophora Root Pills in 120 patients with acne and found to have 73 percent improvement in acne.[34]

A meta-analysis evaluated the therapeutic effect and safety for treatment of acne with acupuncture and moxibustion compared to routine western medicine, and concluded that comprehensive acupuncture-moxibustion was a safe and effective treatment of acne, and possibly better than routine western medicine.[35]

20.7 OTHER THERAPIES

20.7.1 Light Therapy

Acne therapy using various light sources targeting Propionibacterium spp. seems to be promising. The development of infrared non ablative lasers to target sebaceous glands has resulted in the development of a number of laser, light and radiofrequency devices for acne. The main light and laser therapies used to treat acne include intense pulsed light (IPL), pulsed dye lasers, and broad spectrum continuous-wave, blue and red, visible light. A few studies have shown some positive results and light and laser treatments maybe effective and safe for acne but more studies are needed.[36,37]

20.8 CONCLUSION

Preliminary evidence and small pilot studies suggests that Complementary and Alternative Medicine (CAM) may have some value in the treatment of childhood acne. Some emerging data suggests that dietary modification in particular decreasing its glycemic index (GI) and glycemic load (GL), as well as supplementation with ω-3 EFA's may be beneficial in acne management. A few small pilot studies have reported efficacy of some herbs/nutritional supplements, Traditional Chinese Medicine, Ayurvedic herbs and phototherapy in the treatment of childhood acne. However, more research with larger clinical trials is warranted, looking at the effectiveness of CAM in treating childhood acne.

REFERENCES

1. Tom WL, Barrio VR. New insights into adolescent acne. *Curr Opin Pediatr* 2008; 20: 436-440.
2. Eady EA, Gloor M, Leyden JJ. Propionibacterium acnes resistance: a worldwide problem. *Dermatology* 2003; 206: 54-56.
3. Magin P, Pond D, Smith W, *et al*. A systematic review of the evidence for 'myths and misconceptions' in acne management: diet, face-washing and sunlight. *Fam Pract* 2005; 22: 62-70.
4. Davidovici BB, Wolf R. The role of diet in acne: facts and controversies. *Clin Dermatol* 2010; 28: 12-16.
5. Cordain L, Lindeberg S, Hurtado M, *et al*. Acne vulgaris: a disease of Western civilization. *Arch Dermatol* 2002; 138: 1584-1590.
6. Cordain L. Implications for the role of diet in acne. *Semin Cutan Med Surg* 2005; 24: 84-91.
7. Berra B, Rizzo AM. Glycemic index, glycemic load: new evidence for a link with acne. *J Am Coll Nutr* 2009; 28: 450-454.
8. Smith RN, Mann NJ, Braue A, *et al*. A low-glycemic-load diet improves symptoms in acne vulgaris patients: a randomized controlled trial. *Am J Clin Nutr* 2007; 86: 107-115.
9. Adebamowo CA, Spiegelman D, Danby FW, *et al*. High school dietary dairy intake and teenage acne. *J Am Acad Dermatol* 2005; 52: 207-214.
10. Adebamowo CA, Spiegelman D, Berkey C, *et al*. Milk consumption and acne in adolescent girls. *Dermatol Online J* 2006; 12(4): 1.

11. Adebamowo CA, Spiegelman D, Berkey C, *et al*. Milk consumption and acne in teenaged boys. *J Am Acad Dermatol* 2008; 58: 787-793.
12. Deplewski D, Rosenfield RL. Role of hormones in pilosebaceous unit development. *Endocr Rev* 2000; 21: 363-392.
13. Chiu A, Chon SY, Kimball AB. The response of skin disease to stress: changes in the severity of acne vulgaris as affected by examination stress. *Arch Dermatol* 2003; 139: 897-900.
14. Yosipovitch G, Tang M, Dawn AG, *et al*. Study of psychological stress, sebum production and acne vulgaris in adolescents. *Acta Derm Venereol* 2007; 87: 135-139.
15. Zouboulis CC. Is acne vulgaris a genuine inflammatory disease? *Dermatology* 2001; 203: 277-279.
16. Calder PC. Dietary modification of inflammation with lipids. *Proc Nutr Soc* 2002; 61: 345-358.
17. Logan AC. Omega-3 fatty acids and acne. *Arch Dermatol* 2003; 139: 941-942.
18. Yalcin AS. Emerging therapeutic potential of whey proteins and peptides. *Curr Pharm Des* 2006; 12: 1637-1643.
19. Kim J, Ko Y, Park YK, *et al*. Dietary effect of lactoferrin-enriched fermented milk on skin surface lipid and clinical improvement of acne vulgaris. *Nutrition* 2010; 26: 902-909.
20. Mueller EA, Trapp S, Frentzel A, *et al*. Efficacy and tolerability of oral lactoferrin supplementation in mild to moderate acne vulgaris: an exploratory study. *Curr Med Res Opin* 2011; 27: 793-797.
21. Sardana K, Garg VK. An observational study of methionine-bound zinc with antioxidants for mild to moderate acne vulgaris. *Dermatol Ther* 2010; 23: 411-418.
22. Docherty JJ, McEwen HA, Sweet TJ, *et al*. Resveratrol inhibition of Propionibacterium acnes. *J Antimicrob Chemother* 2007; 59: 1182-1184.
23. Fabbrocini G, Staibano S, De Rosa G, *et al*. Resveratrol-containing gel for the treatment of acne vulgaris: a single-blind, vehicle-controlled, pilot study. *Am J Clin Dermatol* 2011; 12: 133-141.
24. Natural Standard. Witch Hazel. http://naturalstandard.com/search-advanced.asp?text=witch+hazel, 2012.
25. National Center for Alternative and Complementary Medicine. Herbs at a Glance: Tea Tree oil. http://nccam.nih.gov/health/tea/treeoil.htm, 2012.
26. Enshaieh S, Jooya A, Siadat AH, *et al*. The efficacy of 5% topical tea tree oil gel in mild to moderate acne vulgaris: a randomized, double-blind placebo-controlled study. *Indian J Dermatol Venereol Leprol* 2007; 73: 22-25.
27. Bassett IB, Pannowitz DL, Barnetson RS. A comparative study of tea-tree oil versus benzoylperoxide in the treatment of acne. *Med J Aust* 1990; 153: 455-458.

28. Elsaie ML, Abdelhamid MF, Elsaaiee LT, *et al.* The efficacy of topical 2% green tea lotion in mild-to-moderate acne vulgaris. *J Drugs Dermatol* 2009; 8: 358-364.
29. Sharquie KE, Al-Turfi IA, Al-Shimary WM. Treatment of acne vulgaris with 2% topical tea lotion. *Saudi Med J* 2006; 27: 83-85.
30. Bowe WP, Shalita AR. Effective over-the-counter acne treatments. *Semin Cutan Med Surg* 2008; 27: 170-176.
31. Lalla JK, Nandedkar SY, Paranjape MH, *et al.* Clinical trials of ayurvedic formulations in the treatment of acne vulgaris. *J Ethnopharmacol* 2001; 78: 99-102.
32. Paranjpe P, Kulkarni PH. Comparative efficacy of four Ayurvedic formulations in the treatment of acne vulgaris: a double-blind randomised placebo-controlled clinical evaluation. *J Ethnopharmacol* 1995; 49: 127-132.
33. Son BK, Yun Y, Choi IH. Efficacy of ah shi point acupuncture on acne vulgaris. *Acupunct Med* 2010; 28: 126-129.
34. Liu W, Shen D, Song P, *et al.* Clinical observation in 86 cases of acne vulgaris treated with Compound Oldenlandis Mixture. *J Tradit Chin Med* 2003; 23: 255-256.
35. Li B, Chai H, Du YH, *et al.* Evaluation of therapeutic effect and safety for clinical randomized and controlled trials of treatment of acne with acupuncture and moxibustion. *Zhongguo Zhen Jiu* 2009; 29: 247-251.
36. Morton CA, Scholefield RD, Whitehurst C, *et al.* An open study to determine the efficacy of blue light in the treatment of mild to moderate acne. *J Dermatolog Treat* 2005; 16: 219-223.
37. Pierard-Franchimont C, Paquet P, Pierard GE. New approaches in light/laser therapies and photodynamic treatment of acne. *Expert Opin Pharmacother* 2011; 12: 493-501.

Chapter 21

Composite Ayurvedic Formulations

Triphala

21.1 INTRODUCTION

Triphala is important poly-herbal formulation of Ayurveda. Work on standardization of poly-herbal formulations used in Ayurveda is in progress and Triphala is one of the few poly-herbal formulations that have been standardized. The standardization of Triphala is based on tannin content.[1]

21.2 COMPOSITION

Triphala is a combination of fruits of three important medicinal plants including *Terminalia chebula*, *Terminalia belerica* and *Emblica officinalis*. Properly dried fruits are mixed in definite proportions to obtain the finished product.[2,3]

21.3 PHARMACOLOGY

21.3.1 Laxative Effect

Gaind, Mittal and Khana reported the purgative (laxative) activity of Triphala in rats.[4]

21.3.2 Radio Protective Effect

Jagetia reported radio protective effect of 0, 5, 6.25, 10, 12.5, 20, 25, 40, 50 and 80 mg/kg body weight of aqueous (water) extract of Triphala. The formulation was given intraperitoneally to animals exposed to gamma radiations. Different doses of Triphala were given for five days. The formulation not only reduced the effect of gamma radiations but delayed the onset of mortality. 10 mg/kg Triphala provided the best protection.[5]

21.3.3 Antimutagenic Effect

Kaur et al., reported the antimutagenic effect of water, chloroform and acetone extracts of Triphala.[6]

21.3.4 Anti-oxidant Effect

Vani et al., reported the anti-oxidant effect of Triphala and its constituents.[7] Naik et al., reported the antioxidant effect of aqueous extract of Triphala. The authors concluded that phenolic compounds present in Triphala are active constituents of Triphala. The anti-oxidant effect cannot be attributed to a single compound.[8]

21.3.5 Cytotoxic Effect

Sandhya et al., reported cytotoxic effect of Triphala in two human breast cancer cell lines.[9]

21.3.6 Immunomodulatory Effect

Srikumar and coworkers demonstrated that oral administration of Triphala stimulates the neutrophil functions in the immunized rats. Triphala also prevented stress- induced suppression in the neutrophil functions.[10]

21.4 DOSE

The dose of powdered Triphala can vary from one teaspoonful (5 g) to three teaspoonfuls (15 g). It is given preferably with warm milk or with warm water.

21.5 STANDARDIZED EXTRACT OF TRIPHALA

Table 21.1 Analytical Specifications of Triphala Extract

Product Description	*Triphala Extract*
Part used	Fruit
Form	Triphala Dry Ext. (4:1)
Appearance	Brown
Odor	Smoky odor
Taste	Sour taste
Solubility	Soluble in organic solvent
Tannin	40 percent +
Heavy Metals	Complies
Microbiological count	Complies

Trikatu

21.6 COMPOSITION

Trikatu is an Ayurvedic poly-herbal formulation consisting of equal amounts of *Piper nigrum, Piper longum* and *Zingiber officinale*. The literal meaning of Trikatu is collection of three acrid drugs. All the three ingredients of Trikatu have appetizer and digestive activity and according to Ayurvedic pharmacology, they enhance each other's activity.[11]

21.7 PHARMACOLOGY

21.7.1 Bioavailability Studies[12]

A. Trikatu and propranolol and theophylline

Bano reported Trikatu reduces the bioavailability of propranolol (beta blocker) and theophylline (bronchodilator) in healthy volunteers.[13]

B. Trikatu and isoniazid

Isoniazid is used in the treatment of tuberculosis. Karan, Bhargava and Garg reported that Trikatu reduces the bioavailability of isoniazid in rabbits.[14]

C. Trikatu and rifampicin

Rifampicin is used in the treatment of tuberculosis. Similar to this work on isoniazid, Karan, Bhargava and Garg reported that Trikatu reduces the bioavailability of rifampicin in rabbits.[15]

D. Trikatu and diclofenac sodium

Lala and coworkers reported that Trikatu reduces the bioavailability of diclofenac sodium.

21.7.2 Hypolipidemic Effect

Kumar and Kumar reported lipid lowering effects of Trikatu in rats fed with high cholesterol diet.[16]

Shilajit

21.8 SYNONYMS

Black Bitumen and Mineral Pitch (English) and *Asphaltum punjabianum* (Latin).

21.9 HISTORY

Ancient Indians were aware about properties of shilajit. Although there is no description of shilajit in the Vedas, Ayurvedic texts like Charaka Samhita have given extensive explanation about shilajit. Shusruta Samhita and Asthangsangreha have indicated that shilajit in the treatment of diabetes insipidus. Charaka has mentioned four varieties of shilajit whereas Sushruta recognizes six varieties of shilajit. Some texts on Indian Alchemy describe two varieties of shilajit:

1. Shilajit having an odor like cow-urine.
2. Shilajit having an odor like camphor.

21.10 DISTRIBUTION[17]

Shilajit is obtained as exudate from steep rocks. This exudate in fact contains 50 percent pure shilajit and the rest are impurities. Some times shilajit is obtained in pure form . Some people describe shilajit as a product of plant and animal origin. Others claim that shilajit is a plant exudate obtained from plants exposed to sunlight. Some consider this plant as *Euphorbia royeleana* and while some describe shilajit as exudates of *Styrax officinalis*. However, the controversy regarding the origin of shilajit is still to be solved.

According to Charaka Samhita shilajit is soft and has light brown color. Charaka further indicates that the best shilajit looks like oleoresin of *Commiphora mukul* (Guggul), has a bitter and pungent taste and smells like cow-urine. Shilajit which is available today varies in consistency from a free-flowing liquid to hard brittle solid. Shilajit present in the market is only stone mixed with mud. It does smell like cow-urine. It is amorphous and colorless. If we compare the ancient and modern description of shilajit it can be said both differ in consistency.

21.11 PROPERTIES

1. Shilajit is semi-solid. It is as thick as honey but soft as compared to confection.
2. Some pharmacies sell shilajit in a crystalline form which becomes moist when exposed to air.
3. Dried shilajit is black in color and smells like cow-urine.
4. Shilajit is a semi-liquid state, yellowish-brown and smells like cow-urine.

21.12 CHEMICAL COMPOSITION

There is a controversy regarding the chemical composition of shilajit. Dr Sidhinandan Mishra in his book on Rasa Shastra has given an excellent comparison of the chemical composition of pure and impure shilajit (Table 21.2).

Table 21.2 Chemical composition of shilajit in the pure and impure

S. No.	*Contents*	*Impure* (percent)	*Pure* (percent)
1	Moisture	12.54	27.03
2	Benzoic acid	6.42	8.58
3	Hippuric acid	5.53	6.13
4	Fatty acids	2.01	1.36
5	Resin and wax	3.28	2.48
6	Gum	15.57	17.32
7	Albuminoids	17.61	16.12
8	Foreign matter	28.52	2.15

(Source: Ayurvedic Rasa Shastra by Sidhinandan Mishra, page 364, Published by Chaukhambha Orientalia, 2nd ed.; 1986).

The constituents of shilajit are of two types:[18,19]

1. **Humic:** They constitute 80-85 percent mass of shilajit. These are produced by interaction of shilajit with lower plants (algae, mosses and liverworts), micro-organisms and even higher plants.
2. **Non-humic:** They constitute 8-10 percent mass of shilajit. Non-humic substance includes low molecular weight chemical compounds (acuparins), oxygenated dibenzo-α-pyrones and triterpene acids of the tirucullane type.

21.13 PHARMACOLOGY[20]

According to Ayurveda, shilajit when administered internally keeping in mind the disease, dose, vehicle and contraindications can cure all diseases. Ayurveda classifies shilajit as *rasyana* (tonic). It can also be taken by a healthy person. This proves that shilajit has preventive and curative properties. Vagbhatta describes shilajit as an anti-aging tonic, aphrodisiac and nootropic. Purified shilajit is used in the treatment of diabetes, fevers, anemia, loss of appetite, colic, obesity, abdominal diseases, spleen diseases, angina pectoris and skin diseases. Medicinal uses of shilajit collected from various sources are described below:

1. **Urinary diseases:** Shilajit is administered with sucrose and honey The dose of shilajit can be from 250 mg to 2 g.
2. Shilajatu Vatika is given in anemia, skin diseases, fevers, spleen diseases, asthma, hemorrhoids, fistula-in-ano, pyuria, seminal diseases, cough, hemorrhage, epistaxis and menorrhagia. The formulation is given with juice of pomegranate or aromatic water. Consumption of *Dolichos lablab* is contraindicated with Shilajatu Vatika.
3. Shilajatvadi Vati is given in spermatorrhoea. It is given with expressed juice of *Parmelia perlata*.

Modern investigations have demonstrated anti-oxidant, anti-anxiety, antidiabetic and aphrodisiac, ulcer healing, hepatoprotective and anti-inflammatory effects of shilajit. A clinical study in Russia has demonstrated efficacy of shilajit in enlarged prostate. Bhattachyra of Banaras Hindu University unlocked the mechanism of antidiabetic action of shilajit. He demonstrated that fulvic acid in shilajit prevents free radical damage to islet of Langherans and beta cells. Fulvic acid (see Fig. 21.1) significantly increased superoxide dismutase activity. Shilajit diminished the development and advancement of diabetes.[21-29]

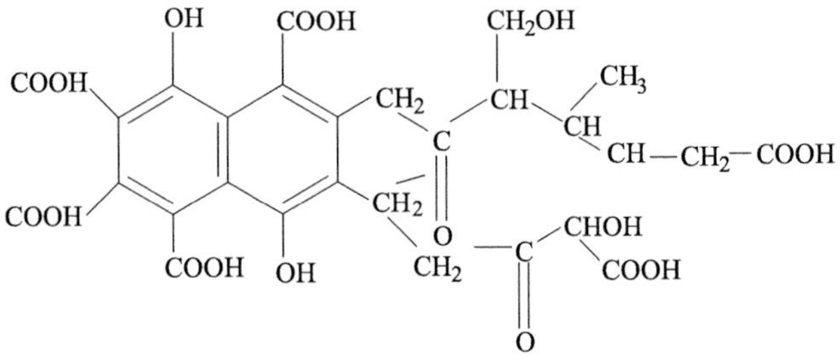

Figure 21.1 Structure of Fulvic acid.

Recent trials conducted in China have demonstrated the usefulness of shilajit in the treatment of diabetic neuropathy. Clinical studies have shown that patients taking shilajit standardized to fulvic acid noted reduction in tingling, painful swelling and numbness associated with diabetic neuropathy. Seeing the promising results of shilajit pharmaceutical use of fulvic acid is approved externally as well as internally in China. Recent studies have shown that systemic administration of standardized extract of shilajit affects cholinergic markers in the brain.[30-31]

REFERENCES

1. Juss SS. Triphala—The Wonder Drug. *Indian Med Gaz* 1997; 131: 194-196.
2. Mahdihassan S. Triphala and its Arabic and Chinese synonyms. *Indian J Hist Sci* 1978; 13: 50-55.
3. Tillotson A, Khalsa KPS. Triphala. *Canadian Journal of Herbalism* 2001; 22: 16-44.
4. Gaind R, Mittal HC, Khana SR. A study on the purgative activity of Triphala. *Indian J Physiol Pharmacol* 1963; 18: 172-175.
5. Jagetia GC, Baliga MS, Malagi KJ, *et al.* The evaluation of the radio protective effect of Triphala (an Ayurvedic rejuvenating drug) in the mice exposed to gamma-radiation. *Phytomedicine* 2002; 9: 99-108.
6. Kaur S, Arora S, Kaur K, *et al.* The *in vitro* antimutagenic activity of Triphala-an Indian herbal drug. *Food Chem Toxicol* 2002; 40: 527-534.
7. Naik GH, Priyadarsini KI, Satav JG, *et al. In vitro* antioxidant studies and free radical reactions of triphala, an ayurvedic formulation and its constituents. *Phytother Res* 2005 13; 19: 582-586.
8. Vani T, Rajani M, Sarkar S, *et al.* Antioxidant properties of the Ayurvedic formulation Triphala and its constituents. *Int J Pharmacog* 1997; 35: 313-317.
9. Sandhya T, Mishra KP. Cytotoxic response of breast cancer cell lines, MCF 7 and T 47 D to triphala and its modification by antioxidants. *Cancer Lett* 2005; 24: 123.
10. Srikumar R, Jeya Partasarathy N, *et al.* Immunomodulatory activity of triphala on neutrophil functions. *Biol Pharm Bull* 2005; 28: 1398-1403.
11. Johri RK, Zutshi U. An Ayurvedic formulation 'Trikatu' and its constituents. *J Ethnopharmacol* 1992; 37: 85-91.
12. Bano G, Raina RK, Zutshi U, *et al.* Effect of piperine on bioavailability and pharmacokinetics of propranolol and theophylline in healthy volunteers. *European J Clin Pharmacol* 1991; 41: 615-617.
13. Karan RS, Bhargava VK, Garg SK. Effect of Trikatu (PIPERINE) on the pharmacokinetic profile of isoniazid in rabbits. *Indian J Pharmacol* 1998; 30: 254-256.
14. Karan RS, Bhargava VK, Garg SK. Effect of trikatu, an Ayurvedic prescription, on the pharmacokinetic profile of rifampicin in rabbits. *J Ethnopharmacol* 1999; 64: 259-264.
15. Lala LG, D' Mello PM, Naik SR. Pharmacokinetic and pharmacodynamic studies on interaction of "Trikatu" with diclofenac sodium. *J Ethnopharmacol* 2004; 91: 277-280.
16. Sivakumar V, Sivakumar S. Effect of an indigenous herbal compound preparation 'Trikatu' on the lipid profiles of atherogenic diet and standard diet fed Rattus norvegicus. *Phytother Res* 2004; 18: 976-981.

17. Ghosal S. *The faces and facts of shilajit*. Proc. National Symp. A Development of Indigenous Drugs in India. New Delhi 1998; 72-80.
18. Ghosal S, Lal J, Singh SK, *et al.* Shilajit: Chemistry of two boactive benzopyrone metabolites. *J Chem Res* 1989; 350-351.
19. Ghosal S, Reddy JP, Lal VK. Shilajit I: chemical constituents. *J Pharm Sci* 1976; 65: 772-773.
20. Acharya SB, Frotan MH, Goel RK, *et al.* Pharmacological Actions of Shilajit. *Indian J Exp Biol* 1988; 26: 775-777.
21. Bhatachyara SK. Activity of Shilajit on alloxan-induced hyperglycemia in rats. *Fitoterapia* 1995; 4: 38.
22. Bhattachrya SK. Shilajit attenuates streptozocin induced diabetes mellitus and decrease in pancreatic islet superoxide dismutase activity in rats. BHU, India.
23. Ghosal S. Chemistry of shilajit, an immunomodulatory Ayurvedic *rasayan*. *Pure & Appl Chem* 1990; 62: 1285-1288.
24. Goel RK, Banerjee RS, Acharya SB. Antiulcerogenic and anti-inflammatory studies with shilajit. *J Ethnopharmacol* 1990; 29: 95-103.
25. Jaiswal AK, Bhattacharya SK. Effects of Shilajit on memory, anxiety and brain monoamines in rats. *Indian J Pharmacol* 1992; 24: 12-17.
26. Saxena N, Dwivedi UN, Singh RK, *et al.* Modulation of oxidative and antioxidative status in diabetes by *Asphaltum panjabinum*. *Diab Care* 2003; 26: 2469-2470.
27. Schliebs R, Liebmann A, Bhattacharya SK, *et al.* Systemic administration of defined extracts from *Withania somnifera* (Indian Ginseng) and Shilajit differentially affects cholinergic but not glutamatergic and GABAergic markers in rat brain. *Neurochem Int* 1997; 30: 181-190.
28. Trivedi NA, Mazumdar B, Bhatt JD, *et al.* Effect of shilajit on blood glucose and lipid profile in alloxan-induced diabetic rats. *Indian J Pharmacol* 2004; 36: 373-376.
29. Vaishwanar I, Koeale CN, Jiddewar GG. Effect of two Ayurvedic drugs Shilajit & Eclinol on changes in liver & serum lipids produced by carbon tetrachloride. *Indian J Exp Biol* 1976; 14: 57-58.
30. Yuan S. Application of Fulvic acid and its derivatives in the fields of agriculture and medicine, Ist ed. June 1993.
31. Schepetkin I, Khlebnikov A, Se Kwon B. Medical drugs from humus matter: Focus on mumie. *Drug Dev Res* 1997; 57: 140-159.

Index

List of Phytochemicals

Z

Index

List of Medicinal Plants

V

Z

Color Plate Section

Chapter 19

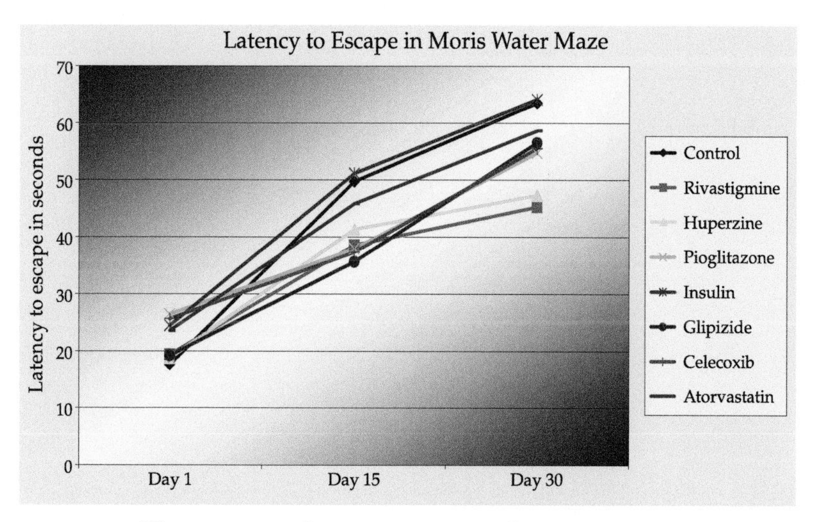

Figure 19.2 See text page **171** for caption.

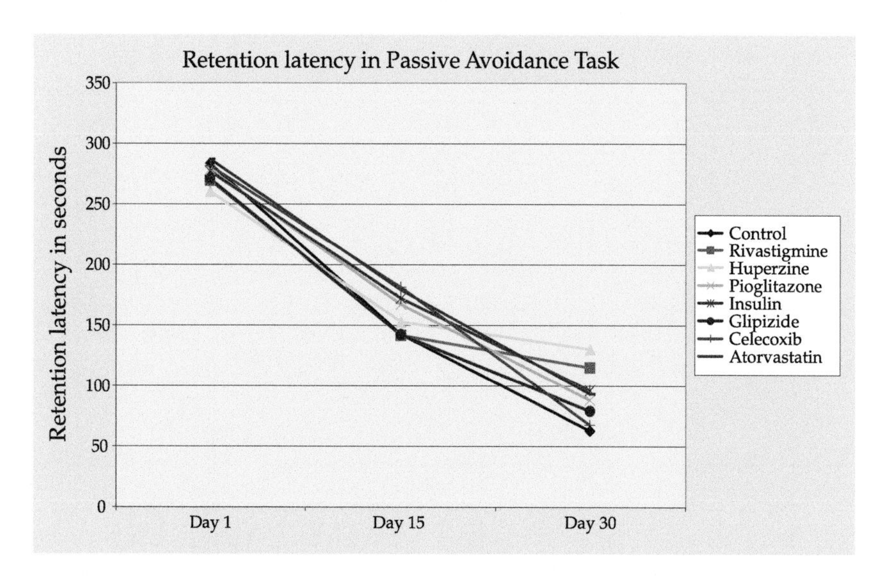

Figure 19.4 See text page **173** for caption.

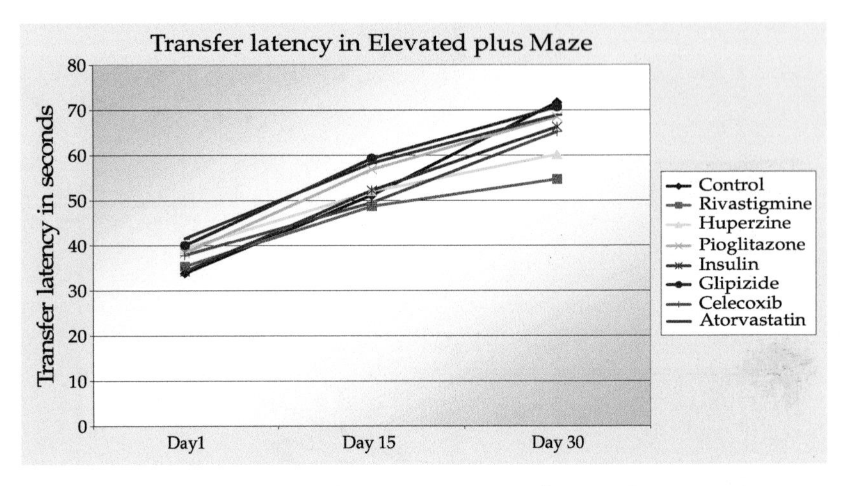

Figure 19.6 See text page **175** for caption.